# AT YONAH MOUNTAIN

## *A Novel*

### *William Stuart Gould M.D.*

WMG LTD.

New York

***WMG***
***WMG Ltd. Publishers***
288 Lexington Avenue
Suite 6-F
New York, New York
10016
wmg.ltd.publishing@gmail.com

For information about WMG Ltd.'s Speakers Bureau or discounts for bulk purchases, please email: wmg.ltd.publishing@gmail.com

ISBN: 0991223799
ISBN: 9780991223794 (pbk)
ISBN: 9780991223701 (ebook)

*Marlene*

*Thank you for the light. I cherish you and it.*

# PREFACE

The night my orders arrived for Ranger School in early 1968, I sought one avenue after another to escape the assignment. Begging for the appointment to be commuted, I besieged friends, and friends of friends, anyone who knew anyone, to intervene on my behalf. I petitioned the Pentagon not to commute my orders for the more pleasant, and, from what I'd heard, safer alternative—a combat assignment in Viet Nam. I moaned to my wife that the reputation of The United States Army Ranger School was so hideous, graduates would not share what they had undergone. Even dropouts blanched at the word Ranger.

Over the past two generations, while swilling countess beers and laughing through my commando tales, friends have asked many times what it would take to get me back to Ranger School, to try it again. "Come on Willy, there's got to be something on Earth worth it. What if the Creator offered wealth, happiness, and perpetual good health for everyone you love? Wouldn't you give Ranger School another shot? Come on."

For nearly forty years, the fleeting thought of even setting foot on one of the military bases that had hosted those unsettling days triggered a recurring dream, a nightmare actually, in which the dread of taking a step back into Ranger School left me sweating, heart pounding, my next days lacking repose.

That reaction lasted until my sixtieth birthday, when I looked up one morning and realized I was no longer a young man. Studying my life critically, I nodded peacefully that, all in all, even with the inevitable ups and downs of life, the years since Ranger School and Viet Nam had been fulfilling and positive. Perhaps the lessons I'd learned from those two disagreeable interludes, by now but blips on the radarscope, had, in fact, been instrumental in my success and happiness.

I wondered, if that was so, might I gather the courage to do something out of the ordinary to earn the right to speak out? Might I relate to the people of the nation I love, a country once again cleaving over war, social tribulations, and shameful political psychosis, that military training is not about the waste of two or three years of a life? Perhaps it is a tough, but valuable, journey along which young citizens develop skills of leadership and teamwork, where they garner a sense of personal pride, and cultivate notions of honor and dignity. It seems these are crucial life skills that are difficult to wrest from a college education or in the work-a-day world.

And so, I awoke on my sixtieth and decided that going back to Ranger School, to redo the initial phase, was to be, though implausible and impossible, the next step in my life. I had kept myself in excellent physical condition, another vital lesson from the military, particularly Ranger School, and felt, knowing what I know now, I could meet the demands of simply tagging along at Harmony Church for three weeks. Little did I realize, as hard as I had sought to avoid being sent to Ranger School in the '60s, I would fight infinitely harder to be allowed to return.

This time, I petitioned old friends, the lieutenants, captains, majors, and colonels with whom I had served a lifetime before, many of whom had risen to the level of general. Though still friends, they were, first and last, pragmatists. No matter how many thousands of dollars I personally spent on medical and psychiatric studies to attest to my physical and psychological ability to keep up with the young troops for three weeks, the answer from the Pentagon came back each time with a denial.

Eventually, one of the generals wrote, "Frankly Bill, they're scared you're gonna drop dead."

Nearing my sixty-second year, it was time to accept that the two years of effort were also just another blip on the radarscope. But then my daughter, a graduate of the United States Naval Academy, and a marine officer who had served in the Gulf twice, told me that one of her naval heroes had uttered, "If there isn't a way, make one."

And so I made a way. But that is another story. First, I want to tell you about what happened at Yonah Mountain.

# CHAPTER 1
# HARMONY CHURCH

The United States Military Academy
West Point, New York
Early 1968

The minister pronounced, "You may kiss the bride," and saber-bearer Second Lieutenant J.W. Weathersby snapped to attention. On command, he and the other bearers brought their blades into salute, raising an arch of tooled silver over the path of the newlyweds. As the only non-West Point graduate in groom Harold Steele's invited complement of saber-bearers, J.W. inconspicuously scrutinized the lieutenant to his left and mirrored that Pointer's practiced movements. Loathe to share with his seven compatriots that he had never before touched a real sword, J.W.'s brow moistened.

The sabers swung upward to form a canopy of gleaming blades over the bride and groom. J.W. closed his eyes and duped himself into believing that, despite the repeated rejections for a congressional appointment to the United States Military Academy, he had finally become a West Pointer. In his reverie, the sword drooped ever so slightly, allowing the tip of his long knife to pierce the bride's lace headpiece. Her crown and veil hurtled to the polished mahogany floor.

J.W.'s eyes opened to the contorted stares of the attending nobility, and to the gasps of their ladies. Even the groom's eyes rolled upward. But through the haze of the goblets of Piper-Heidsieck J.W. had enjoyed at the prenuptial festivities, he was not even sure it was his saber that had launched the lace, and he ignored the faux pas, maintaining his ramrod straight posture. By the time the guests were seated at the reception, the glasses of champagne were overflowing, and J.W. was not positive there had even been a nearly beheaded bride.

After several additional, sparkling tumblers of bubbly, he wobbled into a marble-lined men's room, instantly awed by the eight-foot-high, beveled, leaded glass windows that overlooked West Point's manicured courtyards and parade grounds. Having never toiled on those fields, he stared wistfully through the crystal panes, dazzled by the cut edges prisming the cold, winter sunlight into brilliant primary colors. He was mesmerized by the Gothic structures in which the Grants, MacArthurs, and Eisenhowers had sweat blood during their four years in the Corps of Cadets.

For a moment he was angrily jealous of the groom and his classmates, but the ethanol exerted its dominion, stirring J.W. Weathersby's insides with a pride and happiness more profound than he had ever before experienced. He was wearing the uniform of the United States Army, chosen by his country to defend his family and his nation against the maniacal threat of world communism. He was twenty-two, soon to be a combat hero in Viet Nam, positive he would survive the war and rise to the highest levels of the military. And if he perished in the struggle, so be it. He would die with honor.

An older man took a position at the next urinal. J.W. noticed, out of the corner of his eye, the chest full of decorations that reached not far from the man's jaw. J.W. lifted his eyes slowly to focus on the three, embroidered gold stars twinkling as brightly as the brass fittings on the porcelain sinks. Without opening his fly, the man turned his head toward J.W. and spoke casually, "Major, you know those green epaulettes on your shoulders and the gold piping on your dress blues, well, Major, to be frank, they clash. No combat ribbons on your jacket. You have to be the only field grade officer in the army who's managed to avoid Viet Nam. How'd you do it?"

"Actually, sir, I'm on orders to the 11th Cav. Colonel George S. Patton commanding, sir." The general's face relaxed for a moment. "I'll be darned. I served under Patton's father in the African desert. Where you stationed now?"

"I'm a platoon leader with the groom, 6th Cav, Fort Meade, sir. It's a new unit, sir. Got some growing pains."

"Does it, Major?" The general's face tightened, and his eyes widened. His bushy eyebrows peaked, all but brushing the vaulted ceiling. "I was unaware that military police majors are filling armor lieutenants' slots these days. Guess the war has changed a lot of things."

"Well, to be honest, sir, you see, sir, I am armor, not military police. I left my epaulettes in Baltimore, well actually back at Fort Meade, on the dining room table, sir. I discovered my oversight an hour before the wedding this morning, sir, and had to borrow..."

"You're not a major?"

"No, sir. I'm a lieutenant. But I had no choice, sir. I had to scrounge the epaulettes, sir. You see, sir, I was staying at my mother's house last night in the Bronx, sir, and when I discovered I had forgotten the brass for my uniform, she recommended I call the nearest military installation this morning, and that was right here, sir, the Military Academy, sir. I asked the post operator for, naturally, sir, a lieutenant, sir, but she put me through to a major, sir, Major Norton Bettigole. He's the commander of the military police detachment, and he lent me the epaulettes, sir, because..."

"Lieutenant," the general huffed.

"It's not his fault, sir. It all made good logistical sense, and..."

"At ease, Lieutenant. You mean to tell me that you took it upon yourself to create a new military uniform? The part I don't like is the color, Lieutenant. You're an armor officer. There's a matter of pride when you represent one of the combat arms. You've got gold piping on your uniform, Armor Branch. But that green on your shoulder epaulettes, Military Police Branch! What is that? As far as I know, from my paltry thirty-six years as an officer, our cops don't ride around in tanks." The general turned away from Weathersby and stared through the windows to the Academy parade grounds on which he had slaved nearly four decades earlier. When he turned back to J.W., he took a deep breath and sighed. "So, let's put this together. You took it upon yourself to introduce a new uniform at a West Point wedding, promote yourself by three ranks, and nearly decapitate the bride. And all in less than an hour."

J.W. found himself braced, chin sucked in the way he imagined cadets spent "beast barracks," that first dreadful summer at the Academy. He chose a midpoint on the general's tunic as the focus of his stare, but through the haze of the champagne, he observed that one of the star-encrusted, eagle-head buttons had been sewn on upside down. That brought a smile to J.W.'s face.

As the man scratched Weathersby's name, real rank, and serial number in an olive drab, spiral notebook, J.W. mumbled, "What's he going to do, send me to Viet Nam?" That brought a terrifying thought to J.W.'s mind. "Sir," he whined, "are you going to make me go to Viet Nam with the 6th Cav?"

The general's eyebrows again billowed, and as he strode through the oak door he muttered, "No, I am not."

J.W. was still standing at attention in front of the urinal when the door to the latrine swung open again. He assumed it was for round two and braced more firmly, but a different officer bristled in, a rear admiral this time. The man stopped in his tracks and stared at J.W. before availing himself of the facilities.

⟩⟨

Nearing midnight, three days after the wedding, a telegram arrived at the Weathersbys' quarters.

PRIORITY FROM THE DEPARTMENT OF THE ARMY

J.W.'s wife, Krista, prayed it was a communication forgiving her husband's impetuous decision to volunteer for Viet Nam. In a sense it was.

ORDERS ASSIGNING OFFICER 11TH ARMORED
CAVALRY REGIMENT VIET NAM HEREBY RESCINDED
STOP REPORT WITHIN TWENTY-FOUR HOURS
UNITED STATES ARMY RANGER SCHOOL FORT
BENNING GEORGIA CLASS 68-B STOP IF TRNG NOT
COMPLETED OFFICER WILL REPORT IMMEDIATELY
TO 1ST INFANTRY DIVISION VIETNAM STOP

Krista read the telegram twice then asked, "J.W., what does the First Infantry Division have to do with anything? You're armor branch. I thought your orders were for the 11th Cav. How can they change that to infantry?"

J.W. shrugged numbly. "I donno."

Krista went on. "I keep hearing from the wives about all those infantry platoon leaders getting hurt. And all those infantry officers do is walk in the rice paddies. At least in the Cav, you ride in a tank. Isn't that right, Sweetheart? Can't you get this changed?"

Army Headquarters at the Pentagon was staffed in the wee hours by a duty officer who sizzled, "Lieutenant, call back in the morning at a more appropriate time." The phone dropped with a whack. At zero-seven-hundred hours, the early-morning duty officer, a colonel, speculated that sudden changes in itinerary were usually precipitated by special request. J.W. protested adamantly that he hadn't requested anything but a transfer out of the 6th

Cav, and that he would offer his first-born son to avoid Ranger School.

The colonel remained steadfast. "Are you aware, Lieutenant, of the value of Ranger training in terms of one's career? Only one percent of the officer corps is chosen to attend the commando school. That's a rarified atmosphere. Ever noticed who wears the Ranger Tab? Just about every ranking officer in this man's army. Your file says you're a Distinguished Military Graduate from Sterling College, a pilot—looks like you're regular army. You must've been thinking of a military career at some time. What changed your mind?"

"Nothing, sir. I just want to go to Viet Nam and do my duty, sir."

"That's admirable, Lieutenant, but no changes allowed. I know you've heard all the horror stories about Ranger School, everybody's heard them, but look on the bright side, compared to Baltimore, Georgia's a tropical paradise. Lot warmer down there than up here in the north. Anyway, nothing in this world is as bad as they say it is, right?"

"I guess not, sir."

There was a long pause and then a smirking laugh. "Lieutenant, I hate to tell you this—Ranger School's worse."

"Sir, is there any way I can sign up for the short course?"

"What are you talking about? I've never heard of a short course."

"Sir, that's where you jump off a footlocker and eat a worm."

"Sorry, Charlie, they're waiting for you down there at Harmony Church. Sounds like the training might do you some good. And, you might want to finish the course. I hear the First Infantry Division's running low on platoon leaders out there in the rice paddies. Attrition—eighty to ninety percent."

"Sir, if I graduate from Ranger School, do I still have to go to the First Division?"

"Nope, 11th Cav'll be waitin' for ya with open arms. See, Lieutenant, you might want to keep your mouth shut and play the game."

—=‡+ +‡=—

Late that afternoon, dressed for the Deep South in his summer-weight, short-sleeved, khaki uniform, J.W. shivered uncontrollably as he paced the shoulder of a dusty, frozen, Georgia highway. While the Columbus Airport taxi driver unloaded J.W.'s suitcases from the rusting trunk of the '54 Chevy, Weathersby grumbled, "How long's this gonna take, sir?"

The cabby answered in a slow drawl, "Well, that's part up to you, soldier. Be a lot faster if y'all give me a hand draggin' the spare outta the trunk and changin' this here tire. When we git the wheel fixed, then y'all give the Chivy a push. Hep me git it outta this here ditch. And where in the hell you headed in such a dang hurry?"

"Ranger School."

The driver stood and shook his head. He reached into the cab to turn off the meter. An hour-and-a-half later, the Chevy pinged through the gates of Fort Benning, coming to rest under an enormous billboard imprinted with the black silhouette of an infantryman, arm raised high, the motto FOLLOW ME painted in titanic, light blue letters over his head. A spit-shined guard examined J.W.'s papers then saluted smartly as the car drove off.

"Enjoy that one. May be the last salute y'all be gittin' for a while, soldier," the cabby offered, pokerfaced.

They rolled past clean, new, white barracks, a monstrous PX, theaters, the post gymnasium, and troops lounging at restaurants and playing basketball on outdoor courts. J.W. noticed a sign for the bachelor officer quarters a hundred yards before they approached the red brick structure. He slid toward the cab door, readying himself to be dropped off, avoiding the rearview mirror in embarrassment for having overreacted, for having carped at the driver, and for having groveled at the feet of the Pentagon colonel. But the cabby didn't slow, and they drove several more miles through the Infantry School to the edge of civilization. A weathered sign read:

LEAVING MAIN POST—ROUGH ROAD AHEAD

They jounced along cratered, military roads for several miles until coming to a tumble-down, frame building with a decaying steeple. The cabby mumbled, "Harmony Church—be there presently."

"There's a church way out here?"

"Used to be. These days, that's just the name of this here part of Fort Benning. Was a Negro church fifty years ago, but Uncle Sam bought 'em out. Took the land, too. There ain't no real church here, and there sure as hell ain't no harmony."

The Chevy bumped into a rustic area dotted with pre-World War II barracks and coasted to a stop in front of an eight- by-eight, windowless, wooden shack bare of all decoration, save a freshly-painted replica of the orange and black Ranger Tab. J.W. exited the cab, and facing the hut, noticed a jagged, near-circular patch of washed out, crimson-black paint at eye level on the door's rough wood. It sat directly under the tab. He turned to ask the cabby what it meant, but the driver had already pulled J.W.'s bags from the trunk and was jolting off.

A scowling staff sergeant appeared from behind the shack, tugging up his zipper as he strode toward J.W. He glanced at Weathersby's nametag and roared, "Get your butt into Bravo Barracks." J.W., taken aback by the sergeant's tone, glared irritably at the man, expecting a salute, but the soldier harrumphed and stomped away. He jogged up to another lost soul and screeched in the young troop's ears, "You, Delta Barracks."

A thin, moth-eaten, khaki blanket, a two-inch thick, tattered Gl mattress, and a neat pile of papers lay on the springs of each bed in the unheated buildings. J.W. dropped his suitcases in front of the rack stenciled with "Weathersby." The white paint was still wet. He sat upon the wire springs to read through the life insurance, next of kin, and medical questionnaires. The last card in the heap had but one question: "How-far can you swim?"

A dozen shavetail lieutenants filtered in, found their names, and started on their stacks of forms. They nodded to each other nervously but remained quietly on their own bedsprings. At dusk, the staff sergeant marched into the barracks and ordered the anxious assemblage to fall in back at the stark, wooden shack. Outside, bodies taut, primed for quarry, stood a dozen cadres, officers, and enlisted men, all under cover of black berets bearing fiery orange and jet-black Ranger Tabs.

At precisely eighteen-hundred hours, with J.W. and his 159 colleagues at rigid attention in an unbent queue, stretching, seemingly, halfway back to main post, the first of his classmates was commanded to knock on the rough-hewn door of the hovel. "You will be told when to pass through the portal into the command complex, Ranger."

The first man rapped at the door interminably until a muffled, angry voice from within bellowed, "Enter." The troop remained inside behind closed doors for just seconds then flew out with an expression of terror the likes of which J.W. hadn't seen since passing the mirror before showing his father his report cards.

One by one, each of the cadets made his way to the shack. But every three or four soldiers, there was a delay with a lot of bellowing, and it wasn't until nineteen-thirty hours that J.W. took his turn at the unpainted door. His eyes fixed on the mysterious, crimson-black, central circle. The staff sergeant screamed in his ear, "Knock hard, Ranger. You want them to hear you inside the command center, don't you?"

J.W. struck the door with his fist. That effort, and the next seven or eight, fell short of Harmony Church standards, and by the time he was granted entrance, his hands had been beaten so raw, bubbles of burgundy oozed from the knuckles, his life fluid fusing with that of countless, previous generations of Rangers.

A single, bare, forty-watt light swinging ever so slightly at eye level illuminated the tiny shack. J.W. came to attention, his nose an inch from the bulb. He could make out only the gloom of a human form seated below it. J.W. squinted to focus on the reflection of captain's bars marred by the tiny pits of age and untold encounters with the Brasso rag. He craned his neck for a better view.

*"Ranger, what the hell do you think you are looking at?"*

J.W. shouted, "Nothing, sir!" When there was no further word from the seated figure, J.W. continued, "Lieutenant Weathersby reporting as directed, sir!"

"So, you're Weathersby. We've been alerted you were joining the school," sniffed a pockmarked, pasty-complexioned face. J.W. glanced down to the captain's nametag—VOCK was embroidered above the right pocket. The captain sat behind a two-foot-square, olive drab, field desk. On the wall behind him was a water-stained portrait of Lynden Baines Johnson.

Captain Vock lifted a pile of documents from the desk. J.W. waited to be complimented for his timely arrival in Georgia, but Vock shuffled through the papers silently. When he stifled a yawn, J.W.'s shoulders loosened, but they stiffened abruptly when the man bellowed, "Well, look who we have here. The big football hero. Mister tough guy. Betcha the girls like that." The captain perused a few more forms then barked, "You are not a lieutenant anymore. You are a Ranger. You got that straight?"

"Yes. sir."

"Dismis...wait a minute." Vock had come to the last sheet in the pile. "Says here Ranger Weathersby can swim, what is this, 'FOREVER'? Read this for me, First Sergeant Cowsen. Is that what it says? Maybe I should get glasses."

J.W., though still braced at attention, shifted his eyes imperceptibly to the right. He discerned another soul, a tall, enlisted man with master sergeant's stripes on a Class A uniform. The senior sergeant took a step out of the shadows, and J.W. saw the black man's pleasant face, crisp uniform, and shoulder patch from the First Infantry Division. J.W.'s eyes focused on the rows of combat decorations, several of World War II vintage, some from Korea, and a few from Viet Nam. Though he did not know the significance of most, the two ribbons at the top gave him pause, and J.W. braced more rigidly. Every soldier recognized the Silver Star and the Soldier's Medal, lofty decorations for heroism, the latter for exceptional heroism not in the line of fire. J.W.'s spine straightened even tauter.

Sergeant Cowsen peeked over Captain Vock's shoulder and concurred. "Yes, sir, that's what it says. Man says he can swim forever."

"Nah, couldn't be. Nothing can swim forever, can it. Sergeant Cowsen?"

Cowsen shrugged, "No, sir, nothin' ceptin' a fish or somethin', sir." J.W. allowed silently that there had really had been no reason to antagonize these gentlemen so soon after arriving, but it was, he laughed to himself, water under the bridge.

"That's pretty goddamn funny, Ranger, isn't it? We'll see how many laughs you get at the pool tomorrow. Dismissed!"

As each cadet ran from the shack, he was directed into another hut where a grinning corporal snapped an order. "Drop your butt onto this chair. Remove your hat and do not move." The instant the hat lifted, the soldier swiped a clipper over his hostage's head four times. In eleven seconds, a depilated, embryonic commando was sent back out into the night, guided, not tranquilly, toward the supply shack.

It was bitterly cold at 3:00 A.M. as J.W. and his barrack mates dragged past Harmony Church, faltering along the bombed out road that had delivered them to the Ranger School twelve hours earlier. J.W. paired up informally with a thin troop whose skewed nametag read FRICKER. The two bitched incessantly about the cold, their fatigue, their hunger, their boots and socks, and old girlfriends, until they ran out of breath. Driven at the double-time for much of the march back to main post, few were fully vertical when they hobbled past the BOQ, the PX, and, finally, the basketball courts. The troops were right-turn marched onto Morrison Avenue, where a dozen screeching cadres pushed them to sprint for the final blocks. When a sudden halt was commanded, one-hundred and sixty men accordioned, several falling to the pavement.

Instead of grumbling as they gained their balance, the men grinned, for they had come to rest in front of the Officers' Club. J.W. nodded to his new friend, "Finally, some breakfast. I'm starved." Fricker smiled back and whispered, "Breakfast at the O-Club. I'm going to have oatmeal and English muffins." Though several troops to J.W.'s flanks nodded approval, and a few hooted, there came from the rank directly behind them an angrily whispered, "Oatmeal, my ass."

J.W. turned to meet the sullen eyes of a diminutive black man with a pencil-thin mustache. When J.W. looked harder, the soldier lifted his head arrogantly. J.W. let his eyes drop to the soldier's chest. Instead of the plastic nametag worn by most of the cadets, the letters BRANCH were neatly embroidered above his right breast pocket.

As the company of sweat-drenched, shivering Rangers was goaded across the Officers' Club's carpeted lobby, Fricker whispered again, "Told ya!"

But they jogged past the ballroom, did two laps around the dining room, then passed through the men's dressing room to exit via the polished, mahogany verandah doors to an outdoor, ice-covered swimming pool. First Sergeant Cowsen counted ten men and told them to stand at attention flanking the stairs of the high diving board. With an index finger, he designated a second squad of ten and thundered for them to jump in and use their rifle butts to breach the glacial crust on the pool.

Several of the troops laughed nervously, sure it was a joke, but another sergeant burst from within the club, ran forward like a linebacker, and shouldered two of the men onto the ice. When they surfaced gasping, Cowsen, along with a growing body of smirking cadres, carped at the shivering to flail about with their arms and rifle butts to pulverize the ice and drive it to the center.

With the Officers' Club pool a shaved ice cone, Cowsen mounted the high diving board to address Class 68-B. From a dozen feet above the one-hundred-and-sixty men, he shouted, "Each of you will leap from this here board in full combat gear. You will hold your weapon above your head, tread water for ten minutes, swim to the edge, and climb out. If you drop your piece when you hit the water, you will dive for it, get out of the pool, return to this here diving board, and repeat the whole cotton-pickin' drill until you get it right."

J. W., who had worked his way inconspicuously to the middle of the queue without comment from Cowsen, chided himself for having lain awake all night fearing reprisal for his aquatic braggadocio. He turned to Fricker and joked until his turn on the high board, then leapt with a sneer into the lumpy water. He swam manfully for the requisite ten minutes and then dog paddled to the edge of the pool. As he twisted his face into a haughty grin, Captain Vock appeared from inside a heated cabana, leaned over the edge, and beamed, "Swim forever, Ranger." But the smile slowly degenerated into a look of scorn that chilled J.W.'s bones more than the frigid water. Vock watched J.W. tense in fright then turned and tramped back to the cabana.

J.W. scissor-kicked until the last man passed the swimming test, and then until First Sergeant Cowsen was satisfied that the dripping mass of Rangers braced at attention had fixed their stares unwaveringly upon Weathersby. Captain Vock emerged again from the bathhouse and addressed the company. "Rangers, we are going to be here for a long time, a very long time. The reason I say that is because forever is a long time. Ranger Weathersby can swim forever. He said so. And we're going to wait here until he does."

Vock paced alongside the pool, stopping to shout, "Now, Rangers, we know that Ranger Weathersby can swim forever. Do you know how we know that?" Without waiting for an answer, he went on, "Well, I'll tell you how we know. Because Rangers don't lie." As he and Cowsen ducked into the cabana to warm their hands, Vock called over his shoulder, "Keep swimming, Ranger,"

It was hard for J.W. not to notice the conspicuous lack of charity in the faces of his associates. He drew himself underwater, feigning total collapse, seeking pity, and perhaps a colleague's rescue, but when he resurfaced alive, the disappointed stares had only hardened. He tread water for another minute then swam to the edge and lifted himself out, steeling himself for the order to return to the three-meter board, or worse, to the front of the line that was being reformed to have the entire company scale the tower for another assault on the pool.

As the first man took a furious step forward to mount the high board, Captain Vock stepped from the cabana and bellowed, "Where are you going Ranger? Fall in."

Entropy in reverse, a shattered teacup reforming, a hundred-and-a-half men coalesced instantaneously into four perfectly rectangular platoons of waterlogged, head-shaven, trembling, famished, nation's finest fighting men. J.W. turned his eyes upon the troops to his sides, seeking to ensure that he had melded back into the nameless, faceless mass, but the man on his left jeered, "Thanks a lot, man."

When J.W. turned to his right, that comrade hooted, "Pussy! What the hell you starin' at?"

J.W. squinted into the sneer of the little black man named Branch. He started to explain that his words had been misunderstood by the cadres, that he had only meant to tell them he was a strong swimmer, having learned at an exclusive summer camp in the Catskills Mountains in rural New York State. After two or three words, though, the company of Rangers was driven back into the club to stand at attention. When the carpet was duly saturated, they were moved past the ballroom to do a lap around the dining room, and then onto Morrison Avenue. J.W. raised his hand and asked, "First Sergeant Cowsen, how far is it back to Harmony Church?"

"Same as what it took to get here." He shook his head irritably, hissed under his breath, and hollered the Rangers into a double-time that lasted for the next hour. With boots squishing in rhythm to his cadence, Cowsen ran them at a sprint for a few minutes, then relented and allowed them to jog the flats and speed-walk the hills. When they passed the decrepit remains of Harmony Church, J.W. grasped that it would not be long before they were at the barracks for dry uniforms and the mess hall. A peculiar warmth filled his frozen chest.

Five minutes later, a sliver of the Ranger camp rose over the horizon. Fricker grinned excitedly, "This time I am going to have oatmeal. Toast, too.''

But when half the mess hall was finally visible, Cowsen rotated the formation, still at the double-time, to the south, onto a dirt path away from the camp. They continued for another mile cross country, coming ultimately to a lone set of rotting bleachers deep within the tundra of Fort Benning, Georgia.

After the order to halt, Sergeant Cowsen positioned himself at the front of his dripping, steaming troops. "Rangers, any of you don't like heights?" A dozen hands shot up, J.W.'s the highest. "Good." Cowsen remarked casually, "You Rangers take a seat up there in the balcony." J.W. tried to hide behind another cadet, but Cowsen spied him and pointed toward the highest row. "Up there, Swim-Forever. Top row. Move your butt."

J.W. took a seat in the upper gallery of the bleachers, shivering as he looked down through the rickety floorboards into what he was sure was a two-hundred-foot-deep, craggy ravine directly under the rotting planks. He shifted to his left seeking human contact, but when his shoulder touched Branch, who had slid far to the right to get away from J.W., the small man hissed, "Fuck off, pussy."

Cowsen grunted, "Top row, what's going on up there?"

"Nothing, First Sergeant," Branch answered.

But Branch continued slipping away from J.W., and when he reached the end of the plank, Cowsen shook his head and yelled, "Where you goin' Ranger Branch? Let's get nice and cozy up there. Move yo tail to the right 'til y'all be sittin' in Swim-Forever's lap."

Cowsen approached the lectern, drawing himself into rigid attention as he prepared to deliver his welcoming speech. Though there was a hint of the coming sun, the moon still commanded the sky, and its glow burnished his body as if blue footlights were playing upon it. Most dazzling were the toes of his spit-shined jump boots.

"Rangers," he began, his voice as torpidly penetrating as a Southern Baptist minister's, "you are embarking on a journey many of you won't finish. But welcome anyway. Rangers, our school was founded by General William O. Darby. He was a commando leader during World War II. I had the privilege of serving with this exceptional man.

"Now, the name Ranger was chosen in memory and honor of our Revolutionary War commandos. Those men displayed individual courage, initiative, determination, ruggedness, fighting ability, and achievement.

"Rangers, if you don't want to be here, if you don't want to develop those traits, please say so. If you utter the words, 'I quit,' you will be sent away immediately. Some of you who leave will be assigned to combat units in Viet Nam. You who go to war will find that duty easier than being here. On the other hand, finishing this school is a life insurance policy, 'cause all of you are goin' to Viet Nam sooner or later." His voice rose. "But, if you don't want to be here, we don't want you. You must make that decision by yourself.

The question you must ask yourself every minute, every second of the next months of your lives, is how much are you willing to pay in premiums for that insurance policy?"

He became quiet and stock still, and when the eyes of his audience were as large as the setting moon, he went on, his voice softer, but more penetrating. "Gentlemen, there is such a thing as The Ranger Creed. It says that everyone here volunteered for this training. It says that agreeing to be a Ranger means you acknowledge that your country expects more of you. It says that you will never fail your comrades, that you will never leave a fallen soldier behind. It says that you consent to shoulder more than your share of the task. It says that surrender is not a word in our lexicon, and that a Ranger fights on to the objective no matter the obstacles, even if you are the lone survivor. Gentlemen, this assignment is not for the weak or faint of heart.

"Rangers, the secret to remaining here is simple. We will remind you frequently. It is just two little words: 'Drive On.' When you can't go another step, we will tell you to 'Drive On.' When you are in so much pain you want to quit and die, we will tell you to 'Drive On.' When you are so thirsty your tongue is too thick to draw back into your head, we will tell you to 'Drive On.' Sometimes, we will scream the two little words. Sometimes, we will whisper them.

"Now, Rangers, the only equipment you will need to survive in this here Ranger School is one ear so you can hear the words 'Drive On.'"

Cowsen's expression had become sublime, but it suddenly hardened. "Rangers, some of you are going to be strong leaders in the United States Army. I like strong leaders, 'cause I saw with my own eyes—in three wars—what happens when the weak are allowed to lead men in combat. But the worst was in Viet Nam, where I served in the First Infantry Division." Cowsen paused, the muscles of his face tightening even more. "We had a lot of Ranger dropouts in that unit. Now, I don't give a hoot that they didn't want to pay for the insurance of doin' this here program, and I don't give a hoot that they got themselves blown away. But I do care that they got my men hurt. I don't want weak officers leading my men." He nearly shrieked, "Never again, gentlemen!"

For a moment, he stared into the dawn sky as the gathering rays of sun raised the temperature of Harmony Church one or two precious degrees. "I served in the First Infantry Division in Viet Nam. Had our share of Ranger dropouts. Got a lot of good men killed."

Sergeant Cowsen dropped his eyes from the heavens, drew a deep breath, and went on quietly. "Rangers, we got a lot to cover over the next eight weeks, got no time to waste. When we eat in the mess hall, you will be given forty-five seconds to sit and finish your meal. You notice I say 'sit.' That is because a doctor once told me you should always sit down when you eat, and doctors know what they're talking about, because they're doctors. Rangers, forty-five seconds is plenty of time to eat if you choose your food carefully and correctly.

"Okay, Rangers, I've given you a lot to digest. Go ahead and take a break, but think about what I said. Stand on your seats. Smoke 'em if you got 'em."

The top row rose carefully and extracted packs of plastic-wrapped cigarettes from web belt pouches that had originally held first aid kits. J.W.'s smokes were, however, so wet from his extended swim, he wadded them into a ball and flipped it over his shoulder into the canyon. Before the gob hit bottom, Cowsen had charged up the bleachers, coming to rest so close to Weathersby's lips, J.W. tasted the coffee on the master sergeant's breath.

Cowsen whispered, "Ranger, this ain't the Bronx where you came up. Ranger School ain't no garbage pit. Now get down there and police up the mess you made." Cowsen's face tensed, and J.W. thought the man might punch him, but the sergeant spun about and retreated down the planks to his podium.

J.W. rose to descend into the ravine, first bumming a cigarette from Branch, who grudgingly handed him a Chesterfield. When

J.W. asked for a light, Branch snarled, "Shit, man, you need a kick in the chest to get started?" Irritated, Branch flipped the matches crossly toward J.W., who casually lifted a single hand to catch them. But J.W.'s fingers were so cold, the matches fluttered off his fingertips into the ravine, joining his saturated smokes.

From the podium, a dozen feet below, Cowsen added, "You can bring them up, too, Ranger. And if you don't hurry up, I'm gonna have the cadres hang you from that there rope stretchin' over the pit. That's a special rope, a hangin' rope, Ranger. Now get your butt in high gear, 'causin' your fellow Rangers is goin' to be in the push-up position until you get your work done. And they gonna be the ones to hang y'all from that there rope if'n you keep wastin' their time."

J.W. crawled down the bleachers, bumping into wet, freezing, angry colleagues, fellow Rangers dropping slowly onto their hands and toes to await his return. He pretended to ignore the heated cursing.

At the bottom of the stands, J.W. peered into the chasm, at its madly steep walls, hesitating a second too long, for Cowsen was by his side instantly, the boom of his voice impelling J.W. into the cavern. After allowing his feet to catch up with him at the bottom, J.W. picked at a few items, but when he uncovered a snakeskin and a rock-hard rat carcass, he shot out of the gorge with neither the smokes nor the matches.

Cowsen stood by the lectern, his back to the troops. As J.W. crested the lip of the ravine, Cowsen called over his shoulder for the men to lift out of the push-up position and into their seats. One of the Rangers, a tiny slip of a man who had watched through thick glasses as Weathersby flew back up over the brim, muttered, "Ranger, here." The man offered his own smokes and matches surreptitiously.

J.W. presented himself to Cowsen. "Sergeant, Ranger Weathersby reporting. Mission accomplished, First Sergeant. Here are the cigarettes and matches that fell out of my hand."

Cowsen crumpled the smokes, walked to the edge of the abyss, and threw them in. He pointed to the top row, and J.W. followed his finger unsteadily to the balcony. Cowsen marched back to the bottom row and confronted the small man. "Smith, get your slim butt into that pit and get the smokes that slipped out of my hand. And when you're done, go back and retrieve the trash Ranger Weathersby forgot."

"Yes, First Sergeant."

As Cowsen returned to the lectern and began to speak, J.W. plopped back down next to Branch, who growled in a whisper, "Hey, man, you're bad luck. Go sit somewhere else."

Cowsen stopped lecturing and lifted his head to stare into the balcony, his eyes widening in mock surprise. He raved, "Rangers, I didn't say you two could talk. Branch, Weathersby, both of you, assume the position and gimme fifty up there in the cheap seats. And while you're at it, add one at the end for the Rangers of old."

Head to head, the pair began batting out the first of the nearly half-million push-ups Class 68-B would execute during its tenure in Dixie. At push-up number thirty-five, J.W. looked up contritely at Branch, but all he saw of his colleague was the thick stain around Branch's hat band, the lipid remnant of long-forgotten Rangers who had donned that very same, unwashed cap over the millennia. J.W. laughed quietly, "Your lid needs an oil change, my man."

Branch was silent, but several of the Rangers in the balcony snickered. It cost them several fifty-one-repetition sets of calisthenics. That culled snorting laughter from the troops in the loge seats, though it was not until Smith crept out of the ravine with an impressive array of debris—a slice of frozen toast, a condom carried precariously at the end of a stick, and a huge night crawler—that Cowsen exhorted the front row to drop for fifty and one.

Not placated with Smith's pickings, for it did not harbor the original pack of wadded cigarettes, Cowsen sent the entire company, minus the balcony, into the prone position while Smith descended, yet again, into the valley. When he crawled out nine minutes later, Cowsen shouted, "Okay, Rangers, breakfast. Fall in."

Cowsen let the company relax into an easy march, J.W. believed, to honor the quality of their pre-dawn performance, but after two hundred meters the cadence mutated into the double-time, and Cowsen sprinted to the head of the column, calling incessantly for the next hour:

> Some day—my son—will be—like me.
> He'll run—all day—he'll jump—for his pay.
>      RANGER!
>      RANGER!
>      RANGER!
>      RANGER!
>      RANGER!

"I can't hear you, Rangers."

At first, J.W. was comfortable with the cadence he'd learned in airborne school, but the midnight runs there had been far shorter, and it was not long before he and several of his colleagues began to fall behind. As the burning in his legs built, J.W. became increasingly recalcitrant to the first sergeant's demands to speed up. Soon, several of his better-fed cadets dropped to the ground like so many sacks of sand. These men, it came to pass, were the fortunate sons, a cluster of the pudgy soon to be loaded aboard a trailing quarter-ton truck and drummed out of the course. The remainder of the formation continued at a ragged double-time, eyes to the front, threatened with expulsion if heads turned to catch a last glimpse of the washouts.

When they finally tripped into the Harmony Church camp, Cowsen halted the company in front of a drab, wooden structure. J.W. tried to whisper to Branch that maybe they were finally going to eat, but his voice cracked in its hoarseness, and Branch snapped, "Man, shut up."

Cowsen let out a breath of disbelief as he ordered both of them into the prone position for another set of fifty-and-one push-ups. J.W. concentrated on the red mud inches from his eyes, pretending not to hear Branch's hissed threats. He could not, however, ignore the taunting scent of bacon, toast, and home fries slithering along the soupy ground. As J.W. screamed, "And one for the Rangers of old," on his fifty-first push-up, he jerked his legs under him and sprang back to attention, eyes probing for Cowsen. The first sergeant, though, had vaporized, replaced by the grumpy staff sergeant from the night before.

The lesser non-com stepped to the front of the company. "Rangers, welcome to the best meal in Harmony Church. We sincerely hope you will enjoy the repast we have spent hours slaving to prepare for our new guests. The price of admission is fifteen pull-ups on the bar you see to your front. That's the introductory rate. By noon, it's goin' up. Now, gents, this bar, like the door of that there command shack, has serviced eons of Rangers. You will treat it with respect and add one pull-up at the end for the Rangers of old. Are there any questions?"

A hand was raised, and a hesitant voice called from deep within Second Platoon. "Lieutenant Fricker, sir."

The staff sergeant smiled broadly and cupped his hand behind his ear. "Did you hear something, Rangers?" He winked, grinned warmly, and dashed into ranks, coming to rest a sixteenth of an inch from Flicker's lips. "You see these stripes, Ranger? I work for a living. You don't call me sir; you call me by my first name."

"But I don't know your first name."

"It's Sergeant, you idiot. Do you understand? And you are not a lieutenant any more. You are a Ranger! Do you understand, Ranger, Ranger, Ranger, Ranger?"

"Yes, Sergeant."

"Now," the staff sergeant continued, his voice calmed to a near whisper, "what was your question?"

"I have a bad shoulder. Sergeant. I can't do pull-ups. I have a note from the Main Post dispensary. It's right here in my pocket."

"Okay, no problem. Let's just take a look." Fricker unfolded the soaked paper and handed it submissively to the staff sergeant, who nodded and smiled supportively. "Oh, I'm sorry. I had one of those once." The staff sergeant walked out of ranks back to his command position. "Ranger Fricker, front and center."

Fricker took a martial step forward, heeled to the left, marched to the end of his rank, took a hard right, and approached the staff sergeant, saluting and assuming the position of attention.

"Good, Ranger, that was very impressive. Now do that again and come to attention in front of the pull-up bar. The rest of you gimme fifty and go ahead and add one for the Rangers of old while Ranger Fricker here warms up the bar."

Fricker raised his hand to protest, but the sergeant shook his head then nodded toward the bar. Fricker marched to the bar, sighed, and jumped to grab the pipe, though on pull-up number six, he let go, staggered for a few feet, and collapsed. He rolled about moaning but came to rest on his back, lifeless, as if he had been sprayed with Raid.

"I can't do any more," he whimpered.

The staff sergeant dropped to his knees, placed the doctor's note back in Fricker's breast pocket, then brought his lips to Flicker's ears and shrieked, "Drive on, Ranger! Drive on, damn it!" Tears formed in Fricker's eyes. That brought a simpering smile to the staff sergeant's face and a whoop to his lips. "Drive on, you shit bird!''

Fricker choked back sobs, rose to his feet, ground his teeth, and locked his eyes on the bar. He took a single step toward it, but his head drooped and he turned toward the barracks to gather his things. He was done—just another of the early casualties Cowsen had warned would number in the dozens. The balance of the company took little notice. Fricker, though, made a fateful decision, one that would alter the rest of his life. He hesitated for just a fraction of a second to turn and nod good-bye to J.W. That afforded the staff sergeant an instant to bring his lips back to the young officer's ears. The non-com goaded, "Get your sorry ass up there, Ranger."

Ranger Fricker dropped to his knees and fell prone onto the red clay earth of southern Georgia. A veil of silence descended over Ranger Class 68-B, and then over all of Harmony Church. The clattering of dishes, pots, and eating utensils within the mess hall ceased as well. Even Cowsen was voiceless as he peered through the screened door. The world held its breath waiting for Ranger Fricker to decide if he would do a pull-up or be sent to die in a rice paddy half-a-world away.

A gang of Rangers beating out the latest set of push-ups lifted heads to watch, but a cadre screamed from behind, "That dud's gonna cost the rest of you maggots twenty-five more, and then twenty-five for every thirty seconds he stays off that bar. You hear that, Ranger Fricker?" Fricker rose to his knees like a boxer desperately trying to clear his head before the end of the ten count, but his shoulders wilted in resignation, muddy tears dripping off his cheeks.

One of the lectures in ROTC, though, must have penetrated to the level of Fricker's consciousness, for he ruminated that he was an officer in the United States Army, a member of the one percent chosen to attend Ranger School. That didn't come cheap—everyone knew that. Fricker's jaw tightened again, suddenly appearing to all the world as if he was building momentum to lunge back at the pull-up bar. Smiles crossed the faces of the Ranger cadets, and a sense of pride and acceptance of the ways of the Ranger School slowly built in the ranks. But instead of reversing course and reengaging, Fricker took another step away from the bar, and then another, finally breaking into a staggering jog toward the barracks.

He had taken maybe ten steps when an angry voice rose from the push-up position at the back of Second Platoon. "Drive on,

Ranger. Goddamnit, drive on!" Several more voices percolated from the ground, demanding Fricker reconsider.

The staff sergeant yelled, "Twenty-one seconds, twenty-two."

By twenty-five, the entire company was grunting, "Drive on!"

At twenty-seven seconds, Fricker turned back to the bar. At twenty-nine-plus he began his squat for the leap, but it was not until thirty-and-a-half that his fingers finally touched the bar, and the staff sergeant relaxed in a broad grin, addressing the company of sweating Rangers. "Now, Rangers, there's a lesson to be learned here. That Ranger was late. That's gonna cost you all another twenty-five-and-one. Remember, Rangers are always on time."

Fricker restarted the count from number seven, but the staff sergeant's jaw dropped in disbelief, and he reset the meter to one. Fricker faded again on number six, but this time he held on. The staff sergeant studied the dangling corpse with intense interest for a moment, then screamed, "Nobody move, including you, Ranger Fricker." He ran into the mess hall, came out with a chair, placed it below the pull-up bar, mounted it with a bound, set his lips next to Fricker's ear, and urged, not inaudibly, "Drive on, Ranger!"

Fricker's cheeks were wet with tears streaking through the red powder, war paint, as he sweat through numbers seven to fifteen. He screeched in pain, thrashing through one more for the Rangers of old, though no one was listening. By the time he dropped and crawled to the mess hall, the staff sergeant was back on his chair counseling the next client. Ranger Fricker's life-altering event had already dissolved into nothing more than a puff of dust.

The bar was hot by the time J.W. stepped up. As smooth as chromed steel, subtle concavities had formed at the point where the hands of thousands of Rangers had done millions of pull-ups. J.W. was afforded the chair treatment as well, and when he dropped after number sixteen, twisting his ankle, Cowsen rushed out of the mess hall to greet him.

"Just let me take your arm, First Sergeant. I really think I screwed myself up."

"That happened a long time 'fore you got here, Ranger." Cowsen pointed toward the serving line. "Your cryin' just cost you fifteen seconds. Get your butt on line."

＝＋ ＋＝

Inside, the food was spattering about as if an Oklahoma twister had touched down in Harmony Church. Grub was served by a squad of privates who slapped ladles of wet eggs, grease-saturated bacon, and waterlogged mashed potatoes onto metal trays, much of the fodder ricocheting onto the sweat-soaked uniforms of the famished. J.W. grunted impatiently as a large, blubbery troop ahead of him stopped in his tracks to pick a bubble of runny eggs from the front of his fatigues. J.W. stepped aside, expecting the glob to be flicked to the floor, but the cadet jammed it greedily into his mouth. As J.W. focused on the man's name tag, the troop snipped caustically, "It's Morelotto. Nobody gets it right. And can you believe this shit they're serving up?"

The cooks, though laboring behind raw-wood barriers which hid their faces from the Ranger cadets, had no trouble hearing what their customers had to say about the cuisine. When Morelotto stuck his tray forward at the next station for mashed potatoes, the thump of the spoon striking his plate was so vicious, the tray dropped to the floor.

Cowsen was standing over him before the metal clanged a second time. He barked, "Ranger, we don't waste food here! I already told you how' important chow is." He sucked in an even more robust breath. "What about that don't you understand? Now police up this mess and get to the end of the line and try again." He exhaled noisily then forced a sickly smile. "Clock's a tickin'."

Cowsen turned to J.W. He stared as if bewildered. "What are you waiting for, Ranger? You're wasting precious time. Now you're down to twenty-five seconds. MOVE IT!"

J.W. ran to an open seat, slapped his tray down and was about to start jamming food into his mouth, but was distracted by a chaotic slurping to his left. Branch, face buried in his own tray, was inhaling mashed potatoes. It turned J.W.'s stomach to think he was dining with barnyard animals, and instead of eating, he tried to pry open a container of milk. Snippets of waxed cardboard came loose, but no milk. As he realized he was tugging on the wrong side, Cowsen perched above him and grinned, "Ten seconds. Ranger, nine, eight."

J.W., cured suddenly of his gastrointestinal uneasiness, dropped his face into his tray and sucked in the pool of translucent mashed potatoes, slid a strip of bacon between his lip and upper teeth, grabbed a tiny packet of granulated sugar, and stuffed it, along with the container of milk, into his fatigue pocket just as Cowsen blew a whistle in his ear.

The staff sergeant, who had bristled into the mess hall to supervise the exodus of Rangers, locked his attention on the bulge in J.W.'s pocket. He slammed it with his fist, sending a gush of milk into J.W.'s boot. As Weathersby looked up to curse the sergeant, he caught a glimpse of the big clock over the door. It read six fifty-five A.M.

Outside, the company crept into formation, soon wobbling at the double-time back onto the freezing prairie. J.W. grumbled to Branch, "Now where we going? What the hell do they want from us?" When Branch did not answer, and when the man to J.W.'s right ignored him as well, Weathersby's attention shifted to the burning in his legs and lungs. He slowed to a weak jog and twisted his head to search for cadres, though saw only a gaggle of the less stalwart—some on their knees, some prone on the frozen earth. An olive drab, quarter-ton truck, the meat wagon, materialized out of the mist. As the stragglers were pulled into the bed of the open-air ambulance, J.W. turned to Fricker and hissed, "I can't take this shit any more. I'm goin' on the truck. Comin' with me?"

Fricker nodded, and they slowed to a walk, the second to last declaration of resignation. But as their legs finally ceased all movement, aside from a bit of post-mortem twitching, the truck abruptly broke ranks, the bed stuffed full. It sped past them, making for Harmony Church, its cargo to be fed breakfast's leftovers, discharged from the school, and sent to trudge the paddies of Viet Nam, silage for the infantry war machine.

As J.W. watched the ambulance disappear into the haze, he cursed himself for having failed to fail. A moment later, though, as the first real rays of sun broke over the hills of Georgia, he was braced by the tepid radiance and hobbled weakly back into ranks. Fricker started to run as well, but a minute later, when J.W. weakened, and turned to tell his pal he was finished for good this time, there was a stranger at his side.

J.W. studied the men to his flanks. It was the first time he had seen the unshaven, beaten faces of his compatriots in daylight. He smiled to himself as he spotted Branch directly to his right, the man's wispy mustache glistening with sweat.

Coming to a halt at the bleachers, Cowsen charged, "Face the soldier to your right. He is now your Ranger buddy."

Most of the troops accepted that closest neighbor, comfortably pairing off, shaking hands, introducing themselves to the soul upon whom their survival would depend over the next months. For J.W., that would have been Branch, had the little man not turned and darted away. So J.W. shrugged, laughed defensively, and looked behind him, astonished to see Fricker rejoining the main body. Casually drifting toward his first pal, he stopped when he saw Fricker shaking hands with Morelotto.

J.W. shrugged again and searched in several directions, but each man had paired up. His gut clenched, fearing he'd do the course odd man out, the kid on the playground not chosen by either side. It was only when he turned forward that he spied a man with troubled eyes searching left and right. J.W. smiled, recognizing the petite soldier with thick glasses, the one who'd slipped him the pack of smokes in the pre-dawn hours.

As J.W. approached, the man's face brightened. He stuck out his hand. "Ivan Smith, Johnstown, Pee Ayy. Deer huntin's my favorite thing in the world. What's your game? Water sports?"

J.W. launched into a confabulation, justifying himself for the debacle at the officers' club pool, going on until Cowsen barked, "SEATS!" Smith nodded to J.W., motioning toward the bleachers.

"No, Rangers," Cowsen thundered, "SEATS! HERE!"

With Class 68-B rooted at attention, sweaty butts slowly freezing to the tundra, Cowsen commenced the second piece of his reception opus. "Gentlemen," he began, "and I use the term loosely, welcome to the obstacle course. This is Phase Two of the selection process. You will finish the sequence in eighteen minutes. That means three times around. There is a nice surprise at the end of the final lap for those of you who get that far.

"And, Rangers, you must cross the finish line with your Ranger buddy. Gentlemen, your Ranger buddy is the key to completing this here school. General Darby believed in the 'Me and My Pal' concept of survival. You will march together, sleep together, do guard duty together, train together, and chow down together, assuming you characters can find something to eat. Now, back on your feet and get in file for the obstacle course."

Smith meandered as slowly as he could toward the starting line, tugging at the back of Weathersby's wet fatigue jacket to slow him. "Conserve energy," Smith whispered. "Let the others burn themselves out."

Branch and his buddy, an olive-skinned, high-cheekboned man with a patina of shaved, jet- black hair, fell in behind them. Branch cautioned the dark man just loudly enough for J.W. to hear, "If that asshole gets in your way, kick his butt. He ain't gonna make it, anyway. Too fat. Goddamn loser."

Weathersby ignored Branch and kept dawdling, so Branch huffed angrily and pushed past him. J.W. swung his elbow reflexively, striking Branch in the shoulder.

Branch lashed out a fist, and J.W. Sprang for him, hissing, "Loser, huh? I'll kick your... ass!

The two nearly came together, but Smith pulled J.W. to the ground, and the high-cheekboned man wrenched at Branch, who glared at J.W. and seethed, "You were gonna say nigger, weren't you, *weren't you*, honkey?"

"Fuck you, man. I never used that word in my life. Just stay away from me."

Smith jumped in angrily, admonishing, "Hey, jerkoffs, you're not enemies, they are," pointing to Cowsen and the staff sergeant. "Shake hands, goddamnit."

The two growled like bantam dogs for a few seconds then slapped hands, barely touching fingers. Branch and his buddy muttered to each other and dropped behind J.W. and Smith.

As the first two soldiers in line commenced the first round of the obstacle course, Cowsen shouted, "We lost ten men at the swimming test this morning. Twenty more along the road. We're down to one-thirty. By this time tomorrow, Rangers, we'll be missing some more." Cowsen then disinterestedly continued recording names and times as each pair of Rangers stepped up to the starting line. He barely glanced at the bedraggled men until Smith and J.W. appeared.

Cowsen cocked his head and remarked under his breath, "Lord have mercy. You still here, Bad Weather?"

J.W. sucked in a deep breath to explain, but Cowsen pushed the button on the stopwatch and hissed, "That's fifteen seconds you already done frittered away movin' yo gums, Wet Weather."

A hundred yards through the red slop, J.W. and Smith came to the gallows, a set of thick, wet hemp ropes hanging from wooden crossarms. The two leapt forward, grabbed riggings, and swung most of the way across a three-foot deep pond of crimson-orange mud. But it was not far enough to breach the small lagoon, and they gyrated madly, trying to get the rope swinging harder.

Cowsen shook his head in disbelief. "Rangers, are you morons or somethin'? That there rope been created by the cadres of old. Twist and kick as you might, even if you think you is Tarzan, but Rangers, you will never reach the edge of the pigsty. Rope's too short. We be too smart for a Ranger. Now, let go and give a de-servin' soul the chance to make it into Ranger School."

Smith grunted in a whisper, "Fuck you," and resumed pump-ing. J.W. followed his lead.

Branch screamed irritably from behind, "Stop swinging. Let go, fool!"

But Weathersby and Smith ignored Branch's entreaty and swung once again, though their hands soon tired to the point of failure. They dropped sideways into the mire, slowly resurfacing, enshrouded in the viscous, scarlet, frozen mud of the deep South. A lane grader in pressed fatigues stepped back as they exited the bog. He yammered, "Move it, Rangers. Time's a runnin' out."

Smith and J.W. skidded to the next hurdle, an eight-foot, verti-cal, wooden wall coated with a thick glaze of red clay over which they slowly scraped. Branch waited below, cursing and spitting, then flew over the wall in a single try. He and his Ranger buddy caught up with J.W. and Smith at the third station, a parallel set of two-foot diameter, stainless steel culverts buried several feet under the mud. J.W. and Smith dropped easily into the surface openings, but when they reached the buried section, it became nearly impos-sible to grip the ridges and pull themselves forward. J.W. tried to ne-gotiate the mud-lubricated tubes by flailing arms and legs, but that only lacerated his knees until they bled through his fatigues. Barely able to move, he became hysterical in the enclosed, darkened tube and scratched wildly, miring himself deeper into the residual mud. Exhausted, nearly paralyzed, J.W. ceased floundering, accepting that he was going to die, either in that pipe, or, if someone took pity on him and dragged him free, in the jungle a week later.

Smith, however, emerged from his cylinder, looked around for his new Ranger buddy, nodded to himself, dove into J.W.'s pipe, and pulled him through. They stood face to face, crusted with mud and dirt and sand. While Smith laughed, J.W. babbled incoherently.

Smith's rapture, though, was short-lived. A cadre called to them, "Hey, tube worms, you just pissed away another minute grinnin'."

Smith grabbed J.W.'s fatigue jacket and yanked him at a sprint to the oval cinder track and the quarter-mile run that marked the completion of round one. Branch and his buddy took advantage of Smith's burden to gallop past, arriving at Cowsen's station long before the stragglers. The first sergeant stood in front of a growing cluster of mud-soaked troops, those who had refused to crawl through the tube, those unwilling to bestir the effort to scale the muddy wall, and one man who was so disoriented, he ran the quarter mile in the wrong direction.

Other cadres screamed students' names as they passed Cowsen, who silently checked them off the master list, speaking to no one until J.W. and his buddy crossed the line. Then Cowsen smiled, "You two got less than eleven minutes left for two more circuits. So, take a rest, gentlemen. Y'all ain't got the time to make it. Simple as that."

J.W. breathed a sigh of relief—defeat with honor, he nodded to himself. But Smith looked quickly into Cowsen's eyes and swore, "Hell no! We got more minutes than that!"

Without waiting for a reply, Smith hauled J.W. back to the gallows for round two, a circuit they finished with four minutes to go. Passing Cowsen on the fly, the old sergeant laughed aloud, "I told ya. You two is done. Pack your bags, duds."

But at the gallows on the third trip, with the cadres conspicuously absent, J.W. and Smith followed Branch and his buddy, both of whom ignored the ropes, slogging through the soup on foot. The four then ran around the wall and the tubes, and at the quarter-mile track, with the cadres having retreated to the bleachers, cut across the center.

Cowsen was back at his podium. As the other Rangers struggled around the track for the final lap, a third cadre materialized from the frozen mist and waved them into a file across the chasm from the back of the bleachers. J.W. and Smith arrived at the tail end of the line three seconds before their eighteen minutes expired.

"It's a game of inches and seconds. Don't forget that," Smith breathed in relief. "We're home free. They can't hurt us now."

Over the deep abyss behind the bleachers hung a frayed, horizontal hemp line, the hangin' rope Cowsen had threatened to use on J.W. hours before. It was the only path back to the bleachers. The staff sergeant, the one who had greeted the Rangers the night before, and then permanently deafened Fricker at the pull-up bar, stood, clipboard in hand, on the top rung of the bleachers, facing the gorge.

"Gentlemen," he shouted across to the men bunched together for warmth, "The obstacle course has not been completed. There is yet a task to execute. Rangers, you will monkey climb across this latest obstruction to your completing this course. My end of this rope, as you can see, is anchored, sort of, to a twenty-four-inch granite ledge many feet below me. There is sufficient room to stand there—if you are careful. And a Ranger is always careful. Now, once you get to the ledge, assuming you don't lose your grip, drop, and break your fool necks, as many of you will, you will face the vertical rock wall. Then you will use this cargo net," he pointed at his feet, "to climb back up to my position. The net is suspended from the top of the back of the bleachers here. I hope it does not come loose as you climb, but sometimes it does, especially for the fatter Rangers.

"No matter, you will complete the task of climbing the twenty-five feet in less than two minutes. You will then run down the bleachers and report to First Sergeant Cowsen at his podium. He will be thrilled and delighted to see you, those of you who make it. And you will make it if you climb like hell and always maintain three points of contact—two hands and a foot, two feet and a hand. That is a golden rule, Rangers. I will say it again. When you climb in Ranger School, when you climb anything in Ranger School, whether it is a horizontal rope over a cavern like the one before you, or the vertical cargo net below me, you will always maintain three points of contact. We hate it when Rangers fall into the valley you see below you. Lotta paperwork. On the other hand, it is another technique we use to whittle down numbers. Now, first man... BEGIN!"

With that declaration, the staff sergeant grabbed the edge of his clipboard and hurled it across the gorge. It flew like a discus, a Sandy Koufax pitch, curving left and right, up and down, until arriving at the gaggle of mudded Rangers. The more alert troops followed the rocket with their eyes, though Fricker, still mumbling about enjoying a bowl of hot, buttered oatmeal at lunch, was staring down at his feet when the missile caught him in the back of his head. The clunk was audible, so loud in fact, Sergeant Cowsen charged up the bleachers and shook his head sadly as he stared across at Fricker's rumpled form writhing in the mud. Cowsen breathed deeply and hollered forlornly, "Rangers, you must always pay attention." With that pearl of wisdom, he spun around and ran back down to his podium.

The staff sergeant screamed, "MOVE IT!" but when the first pair of Rangers marched up to the rope and hesitated, the sergeant launched himself off the top of the bleachers, down the cargo net, onto the rope, and across the ravine in less than eight seconds. The Rangers assumed he had arrived to address Fricker's wound, but the staff sergeant simply pushed the first man onto the rope, waited until the quaking troop was halfway across, then jumped back on the rope himself and pumped his legs up and down until the single strand of thin line oscillated like the Bridge of San Louis Rey.

By this point, Fricker was back on his feet with his first aid dressing pushed against the spurting gash in his head. Morelotto, Fricker's Ranger buddy, saw the gush of blood, gagged, and was on his knees occupying the spot from which Fricker had risen. Branch took a step back to stand over him, snorting, "Guess you ain't a candidate for a profession in the surgical arts, are you, asshole? Get up and cross the goddamn gulch. I ain't got all day.".

The swarm of Rangers may have heard Branch, and even agreed with him, but there was not a millimeter's movement toward the rope. J.W. grabbed Smith by the shoulder and gasped weakly. "Man, I can't deal with this. I'm afraid of heights. I'm not going across."

"I don't like 'em either, but we can do it. I learned how when I had to go down mountains to get deer we bagged in the Alleghenies. Watch me," Smith demanded. "It's a piece of cake. Just don't look down, man. Always look up. Like I said, watch how I do it."

Smith dragged J.W. by the sleeve to the edge of the ravine. When J.W. looked down, Smith screamed, "Watch me, I said!" Smith spit onto the ground haughtily then dove onto the sagging rope, his eyes, though, glued to the heavens. At first, his slight body vibrated parallel to the Earth, arms and legs wrapped tightly on the prickly nylon. He glided toward the low point of the rope smoothly. At the bottom of the sag, Smith laughed and craned his neck so far back, it looked as if he was locked in a seizure. Eyeing J.W. drooling with fear, staring into the void, he enjoined, "Ranger, I told you to look at me!"

J.W. raised his eyes out of the crater, and as Smith pulled himself along the upswing toward the ledge, he called to Weathersby. "Hey, man, it's fun. Don't worry. Pretend you're going after that eight-point buck you just bagged. Follow me!"

J.W. was about to grasp the rope, but he hesitated for an instant, and Branch hissed, "That's it. Get outta my way." He and his buddy jumped ahead of J.W., but when Branch stared into the chasm, he hesitated just long enough for another set of Ranger buddies to step up, the front one shoving Branch out of the way. Branch elbowed back, and the two pairs of Ranger buddies were soon engaged in animated jostling. It was, though, just so much Brownian motion, for no one dropped onto the rope.

The staff sergeant, who had long since resumed his position atop the bleachers, was apoplectic. He charged down the cargo net onto the cable, scaled over Smith, and vaulted off the far end into the mass of fossilized Rangers. He shrieked, "Rangers, get on the goddamn rope. Now!" And back he flew, snaking over Smith again, onto the granite ledge, where he parked himself, glaring at his apprentice commandos.

Smith hollered up from the rope, "Hey, Sergeant, gimme a minute before anyone else gets on, huh? Just a minute?"

Smith saw the staff sergeant's face redden to the color of a glowing machine gun barrel, and he sped the last few feet. When he stuck his hand out for a boost onto the tiny ledge, the sergeant spit, leapt up the cargo net like a chimpanzee, and shrieked, "Driiiiiiiiiiiiive on, Ranger!"

With a proud laugh. Smith sprang onto the net, a man possessed, yelling back to J.W., "Come on, Ranger. Nothing to it. Just don't look down, man."

An M-14 rifle strapped across his back, held there with a leather bootlace, a Ranger sling, Smith crawled vertically, hand over hand, up the thick hemp of the net. Ten feet from the summit, with the staff sergeant on the balcony throwing small rocks onto his helmet, Smith stopped to suck in a deep breath before the final assault.

The staff sergeant smirked down and called to him, "Too late for your sorry ass! Time's up."

Smith lifted his left hand off the net, raised a perfect bird, and while aiming it at the staff sergeant's face, took a mighty step upward. With his middle finger still wagging, Smith pulled his right hand off the net to grasp at the next rung. With that deed, Ivan Smith of Johnstown, Pennsylvania, became the first Ranger of

Class 68-B, but not the last, to violate the golden rule, the three-points-of-contact commandment.

His spare body peeled slowly backwards, away from the net. For a long moment, it appeared as though he was suspended horizontally, perpendicular to the side of the mountain, as he had been seconds before on the gorge-crossing rope. Inevitably, gravity exerted its fist, and Smith's body was drawn toward the valley floor miles below. At first he fell slowly, arms flapping, grasping in slow motion at the net for a chance to undo his critical mistake. But soon an acceleration of thirty- two feet per second per second, certainly a number he had learned by heart as a mechanical engineer at Carnegie Mellon, declared itself, and he sailed more and more speedily downward

To the men observing in utter silence on the far bank, jaws slack, though, it seemed an hour before the olive drab projectile quickened toward the bottom. And they remained stock-still as Smith bounced to rest on the two-foot-wide ledge thirty feet below the top of the bleachers. His journey terminated with a thud and a loud crack, as if a green bough had snapped. When the recoil and shuddering was over, all that stirred was a splintered wedge of polished mahogany rifle butt, swinging like an upside down metronome in cadence with Smith's moaning.

No one moved or spoke until J.W. Weathersby broke the silence, crying out, "That's my Ranger buddy!" He shoved his way forward, dropped onto the rope, and closed his eyes. They remained shut tightly as he slithered across, arms and legs flailing on the rough hawser. He pulled himself onto the insubstantial ledge, barely finding room to kneel as he took his first look at the flickerless, ashen face that drooled blood from eyes, nose, and ears. J.W. froze, transfixed not by the trauma he was witnessing, but by the shattered rifle butt, and the acknowledgment that, dead or alive, Smith would soon have to answer for the destruction of his army weapon, a blunder that often ended an officer's career. J.W.'s chest gripped even more uncomfortably when he considered Sergeant Cowsen's dictum about the responsibility shared by Ranger buddies, the precarious liability, he muttered to himself, recalling the pre-law class he had endured at Sterling College. That warning still ringing in his ears, J.W. wondered if his future was also dangling by a shoestring over a bottomless pit.

Smith's eyes fluttered. He raised his head slightly and pointed tremulously toward the tiny, Gl issue, first aid pouch on his web belt. "Man, you gotta stop the bleeding," he pleaded. When J.W. reached for his own pack of bandages, Smith whispered, "No, mine first. Save yours," but when J.W opened Smith's pouch, it was devoid of both first aid bandages and the pack of cigarettes that had been sacrificed hours before.

His Ranger buddy of twenty-three minutes lay dying, and J.W. made the brave decision that Smith's life was worth breaking a cardinal rule. He went for his own first aid pouch. It was empty. The first aid packet that had begun Ranger School in his pouch had been tossed away along some country road hours before in a feeble effort to lighten his load, which it had by less than an ounce. The smokes that had taken its place were also on the floor of the ravine, and J.W. looked away from Smith to search for them. He saw both crumpled packets far below on the earth beside the snakeskin and the dead rat. He removed his own tee shirt to tamponade the spurting vessels of Smith's face and scalp, though, despite his efforts, the man's breathing became shallow, and his eyes slowly closed.

From atop the bleachers, the staff sergeant glowered and soon howled, "Move your asses, Rangers! Tell that son of a bitch to get off his butt and climb."

He was joking, of course, J.W. sought to convince himself, but when the staff sergeant screamed louder, "Get off your ass, Ranger!" Weathersby's middle finger unfurled to attention, mimicking the aplomb with which his fallen comrade had saluted the staff sergeant just seconds, and a lifetime, before.

J.W. accompanied his bird with a panicky, "He's hurt. He can't move. I think he's dying. Get a goddamn doctor."

The staff sergeant straightened his spine, grinned, and addressed his audience, chortling to the column of besilenced soldiers, men barely out of their teens, one-hundred-and-thirty of them paused in frozen bewilderment on the far side of the canyon. "That'll just show ya. Rangers don't bounce!"

J.W. stayed beside Smith despite the staff sergeant's brayed instructions to mount the cargo net and get on with the training. Noting the lack of J.W.'s obedience, and the looks of confusion on the men across the gorge, the sergeant launched into an increasingly strident tirade, demanding Weathersby abandon his vigil and mount the net. But J.W. didn't, and after half-an-hour, a stretcher was lowered and two captains edged toward J.W. and Smith, both officers anchored to the mountain by heavy ropes. One captain spoke softly, "We'll take it from here, Ranger. Go on up and join your company."

J.W. mounted the net, grinding through each step, gripping the bristled mesh with three very sure points of contact, resolving to stay alive long enough to summit the obstacle and then abandon the course. At the balcony, he touched the seat he had warmed for so many hours, stopping to contemplate just how to tell the staff sergeant and Cowsen that he'd had enough. But a kick in the ass from the man behind him sent him, arms thrashing, stumbling down toward Cowsen. He came to rest after banging into the lectern, knocking Cowsen's clipboard to the ground. Cowsen's glare so jolted J.W., he squeaked, "Ranger Weathersby has successfully completed the obstacle course."

Smith was gone and J.W. was sure that he, too, was history. He stood at attention with what was left of his platoon, singled out as the focus of the staff sergeant's glare, waiting for the clap on the shoulder pulling him from formation. He played in his mind how he would feel as the cold orders came to fall in alongside the losers, and he wondered if the lasting misery of the incident would be relief or shame. Without much rumination, he knew there would be deep remorse, but a moment later he smiled at how blissful that shame would be.

As the last of the washouts was called by name and instructed to mount the deuce-and-a-half truck for the ride back to Harmony Church, J.W. gathered, with some disappointment, that he had made another cut. As the bedraggled company marched back into the tundra, what overwhelmed him more than the sight of Smith's broken body was that eighteen hours before, both of them had been at home, warm, safe.

The hand did not fall that morning, his first as a commando conscript, and he dragged along during the forced march back to Harmony Church, a thrash that terminated outside the mess hall. The company remained at attention for an hour, their mission, to wait for the next inventory of those who had made the cut. They spent the time batting out sporadic cycles of push-ups for infractions ranging from wrinkled, muddy uniforms to dirty fingernails. Eventually, the list was posted on the command shack door, though the names were not catalogued in the usual fashion. While there was, indeed, a typed register of those who had failed, the names were in haphazard order, badly misspelled, and only a few capitalized.

The list was so long, by the time he found a name that might have been his, the cadres had commenced a summons to fall in, and Cowsen screamed, "Read Forever, what is your problem? Get your sorry bee-hind into formation."

The men whose names were clearly on the list, and a few who pretended to have spotted their names, found themselves bullied into a mini-formation before being marched off, staring at the ground, to accomplish their last act as Rangers—consume a cold brunch of mashed potatoes and bacon left over from the pre-dawn meal.

J.W. watched them trudge toward the barracks, but, with a last minute left-face march, and then a right-face march, they were ordered to a halt in front of the dilapidated mess hall. J.W. could see their heads and shoulders rise; several even cheered when they walked past the pull up bar without a word from the buck sergeant at the head of the column.

Branch laughed sardonically, "Man, every one of them suckers ate too much this morning. Loaded their bellies with so much lead, they couldn't move. Now they're, payin' for their moment of pleasure. Look at 'em grinning—Grateful Duds."

Fricker, who had wrapped his head in his tee shirt to stifle the bleeding from the staff sergeant's clipboard, held a different view: He suggested, "No, man, they're the smart ones. Bein' rewarded twice. Got to eat this morning—gettin' to go home now."

Branch yapped, "If you hate it so much, why the hell don't you just shut up and quit?" Though Fricker pretended not to hear, J.W. glanced quickly at Fricker's eyes and saw the red returning. With that, he punched Fricker in the arm good-naturedly, advising not softly, "Don't listen to 'im. That chip on my man's shoulder over there, it'll bury 'im. Wait and see. He ain't goin' anywhere. Too much baggage."

Branch stood and took a step toward J.W., but the little man's Ranger buddy, the dark man with the chiseled cheek bones, half-stood and wordlessly stretched an arm to stop him. Branch spit on the ground and hissed, "You watch your ass, white boy. It's gonna happen, and just when you ain't lookin'."

For the first hours of the afternoon, following a meal that had been promised, but in the end not extended, Ranger Class 68-B *was* sent to bone up on the finer points of the art and science of beating the living crap out of each other within the confines of a sumo ring. This was Pugil 101, a class they'd first studied in Basic Training. There were plenty of padded sticks in the heap outside the mess hall, but not enough of the archaic, leather football helmets to go around. Cowsen ripped one out of Morelotto's fist and jammed it at J.W. "Here, put it on. You scare me, Ranger," he muttered. "A little more brain damage, and y'all gonna believe y'all can fly forever."

Despite Cowsen's words, J.W. had fretted all afternoon that it would not be long before his presence was challenged, that Cowsen would soon discover the gaff of not having included his name on the list of washouts. The first sergeant would dismiss him publicly, requesting from the remaining Rangers a moment of prayer for the enlisted men of the First Infantry Division, who would soon find themselves under the leadership of Lieutenant J.W. Weathersby, Ranger Drop Out. By midafternoon, however, after hours of verbal maltreatment from Cowsen and the staff sergeant, Fricker began to taunt J.W. as well, "Hey, pal. I got bad news for ya. Looks like you're gonna be a Ranger."

J.W., moved by the notion of acceptance, became desperate to suffer through more so they would allow him to suffer through more. He fought his way down to the last man in the pugil pit, the final opponent, Branch.

Cowsen spoke to the two finalists. "Prize for winning's an extra thirty seconds at the evening meal. Now, both of y'all. I wanna hear y'all growl like a pissed off mama brown bear. Okay, have at it."

Branch yowled and lurched forward just as J.W. opened his mouth to howl. Branch thumped him in the solar plexus, and J.W. doubled over. Branch pummeled madly until J.W. took a blow to the temple and collapsed. The blistering, freezing, Georgia sun vibrated in circles in his battered brain while Branch danced disdainfully in the winner's circle until a hornet stung him on the arm. Branch launched into a dance of overdone agony, chin strap undone, more a black Knute Rockne than the Ranger heavy weight champion of the world.

A cadet in horn-rimmed glasses turned his head in query. "Vespidae don't generally fly during winter. Very curious."

Rather than cheers for having prevailed, Branch's compensation was a distended welt above his wrist. The first sergeant counselled, "Ranger, pain is just weakness leaving the body. Now, fall in and start marching."

J.W. straggled behind the groaning troops who had been dispatched to move at the double-time, bayonets fixed, toward the next tutorial. Along the road, one of the Rangers tripped on an old C-ration can and fell forward, his bayonet stabbing the ankle of the troop to his front. J.W. stopped to help, again trying to stifle the bleeding of one of his comrades, but seeing the staff sergeant bringing up the rear, ran off to rejoin the main contingent. In the distance, J.W. heard the staff sergeant shouting over Cowsen's cadence, "Get off your lazy ass, Ranger. Drive on, Ranger! Drive on, damn it, your ankle's four feet from your heart!"

After the bayonet course, the company was pushed at a run back to the mess hall for pull-ups and forty-five seconds of watered-down mashed potatoes, cold bacon, and warm milk. Following the after-meal pull-ups, they moved at the double-time to a new set of bleachers. Fricker had developed a technique for carrying his right arm close to his body to stabilize the troubled shoulder. He waddled as he ran. The troop who had been stabbed in the ankle dragged his leg—the staff sergeant bellowed, "Hey, Frankenstein, move your slim ass,"—and Branch's arm had swelled to twice its usual size, forcing him to hold it above his head. Cowsen kept asking, "Ranger Branch, do you have a question or a comment?" An hour passed and Branch's swelling crept from his arm into his face. Cowsen halted the troops and stepped up to Branch. "Ranger, you need to get them lips under control."

At the bleachers, the cadres spoke for two hours, delivering an articulate critique of their students' first day of training. The staff sergeant was the last to deliver his message. "You Rangers who think this is some kinda vacation ain't gonna be here long. I'm here to tell ya'; this ain't no holiday."

At 10 P.M., the company force-marched back to Harmony Church. Most of the troops dropped on the nearest rack in what was left of their filthy, shredded uniforms, though a few undressed and washed their hands and faces. One or two brushed their teeth. Some of the men talked for a while before closing their eyes, and Branch and Kenyon, two of the very few black Ranger cadets, spoke in whispers, laughing quietly.

"Man, I came up in Spanish Harlem," Branch divulged. "I don't know shit about 'dem woods out there, my man."

Kenyon laughed, "Man, I'm city too, Philly, but I figure I know about jungles, man. Can't be that different. Urban concentration camps, man. You know what I'm sayin'?"

J.W. sat up in his rack and called quietly to Branch and Kenyon, "Hey, man, I came up in the city too, the Bronx. I don't know nothing 'bout the forest either. I know 'bout the city though—never stare at no one. I never look down when I'm on the street, always straight ahead, hands out of my pockets, not hidin' a piece."

Kenyon nodded politely then whispered, "I can dig it."

But Branch did not answer. Muttering under his breath, he left his rack to go outside for a smoke.

In seconds, loud snoring and the rustling of beaten men resonated with the boot steps of First Sergeant Cowsen, who had ambled through the back door into the barracks. Those who were awake jumped off their beds and dragged themselves to attention, but he told them softly, "Get back into your racks, gentlemen."

He went to the side of J.W.'s bunk and spoke coldly. "Ranger, report to the command shack. Captain Vock wants to see you."

And so the series of mistakes that had kept J.W. Weathersby in Ranger School for a full day had finally been unearthed. He pulled out the last clean uniform from his footlocker and his second pair of boots, still gleaming from the daily spit-shine back at Fort Meade. He checked his gig line for the straight arrangement of his fatigue shirt hem, the edge of his brass belt buckle, and the flap over the fly on his fatigue bottoms. No use in antagonizing the cadres any more than he already had.

Weathersby knocked so hard on the wooden door of the command shack, the superficial scabs on his knuckles rubbed off, and though he could feel the blood oozing from his hand, it was too dark outside to appreciate how much he had just added to the burgundy patch.

Vock sat at his field desk. "Do you have to make so much damn noise when you knock, Ranger?" Before J.W. could answer, Vock went on. "That was very brave of you today, Ranger. Wasn't it?"

"What is that, sir?"

"That little episode with Ranger Smith."

"Just doing my job, sir, sticking by my Ranger buddy."

"Well, the commandant thought it was hot shit. That's why you did it, isn't it?

"Sir?"

"To impress somebody and save your butt?"

"No, sir. You told us to treat our Ranger buddy like he was more important than we were…"

"My ass. You dodged the bullet this time, Ranger, but I promise you, one more episode, and you will pay. I awarded you *minus* twenty-five points for disobeying a direct order from the staff sergeant, but the commandant gave them back to you, and then gave you twenty-five more for some unknown reason. But screw up again, Ranger, and your ass'll be on a Flying Tigers Airways tub to Viet Nam so fast, the blood on your knuckles won't have had time to clot. Am I understood?"

Before J.W. could answer, Cowsen opened the door to the shed and spit, "And may the good Lord help the men of the First Division when you're put in command of an infantry platoon."

The captain shook his head then looked down, writing in J.W.'s personnel file. When J.W. did not move, Vock blurted without looking up, "Dismissed."

As J.W. lay in his rack, he turned the day over in his mind, the thousand snares barely eluded, the relief over having survived, and the disappointment and fear of having done so. But it was the discussion with Vock that kept him from drifting off to sleep. He guessed from the captain's words that he had somehow made the cut, and that was good, but his gut soured as he contemplated how his presence could already have been made known to the commandant. He also thought back to Vock's welcoming words, that he had been alerted J.W. was joining the class. His gut clenched once again, though relaxed a moment later as he reassured himself that was what Vock told everyone. Of course it was.

He looked up and saw Branch tossing about in bed. J.W. called softly, "Hey, man, did they tell you last night they knew you were coming?"

"What the hell you talkin' 'bout? Go to sleep, fool."

As fast as the barracks lights were extinguished, they flickered back on. J.W. and Branch were in adjoining bunks, lying on the bare, rusted, metal springs, too tired to have unfolded their moth-eaten, inch-thick GI mattresses. The soldier who had been stabbed in the ankle was on the bunk above J.W., his puncture wound wrapped in his mattress, though the blood had seeped through and a few drops had trickled down onto J.W.'s face. J.W. felt the blood and rubbed it off with his finger. He gasped, but soon calmed, believing it to have come from his own battered knuckles.

A voice hollered from another upper bunk, "Hey, turn off the goddamn light, or I'll kick your ass into the next fuckin' time zone," but the lights stayed on. J.W. looked up; Cowsen had one hand on the switch, the other held a thin stack of white envelopes.

"Good morning, gentlemen. Off your butts and on your feet." Most of the men tore for the door, but Cowsen screamed, "Where you goin'? Stand in front of your bunks!"

Those who had taken the time to get out of their fatigues a few hours before were in their drawers, all except Fricker, who had slept in the buff. He still had toothpaste on his chin.

Cowsen walked to the center of the room and barked, "The new company mailman." He flipped the envelopes at Fricker, who was caught unaware, struggling to pull on a pair of drawers. The boxers fell to the floor as he grabbed to catch the packet of mail. He shrugged indifferently, stripped the away the rubber band, and began calling names. "Bearchild!" Branch's Ranger buddy, his onyx eyes gleaming, stepped forward, nodded, and accepted an envelope "Branch!"

"Yo!"

"Weathersby!"

"Yo!"

Branch ogled Weathersby's envelope, laughing contemptuously at the scribbled, secret messages and the extra, upside-down stamps Krista had stuck on to post it airmail, special delivery. J.W. sneered back and headed for the latrine, seeking a moment's peace to peace to read the letter, but Cowsen had slipped outside the barracks and blustered, "Fall in. You got three minutes. You will be freshly shaved and presentable. You are soldiers, and more important, a few of you will be Rangers someday, maybe."

J.W.'s fellows ran past him into the latrine, splashing water on their faces at a row of communal sinks. Those who could not squeeze in dashed from the john to dry-shave by their bunks. By the time J.W. ran back to his footlocker for a razor, and then back to the sinks, he was the lone soul in the latrine. He sprinted into the barracks proper. It, too, was empty. Unshaven, only half-dressed, he ran frantically to the window and peeped through a streak in the frost-encrusted glass. A tight formation of his comrades stood weaving in the bitterly cold, pre-dawn mist.

He heard Cowsen. "Welcome to Day Two of Ranger School. Rangers, ten hut. Left face. Some of you will still be here tonight. Some will not. At the double-time, forward march." For J.W. Weathersby, that moment of hesitation, that six-tenths of a second he stood immobilized at the window, cast a die. He was done, the rules of the game of inches and seconds Smith had characterized were about to hang him. He could not fathom how it could have happened that fast.

The heavy, leather boots of his compatriots pounded into the distance, Cowsen roaring cadence. J.W., who had ducked to hide under the window, rose and peered out as the company jogged away from Harmony Church, sparkling crystals of frozen mist settling on the stubble of what was left of their hair. As the beat faded, J.W., utterly alone in a barracks for the first time, drifted in a quaver toward the latrine. His chest was so knotted, he dropped onto the center throne to think, but just stared at the row of ancient, cold-water-only sinks. His hand brushed his pocket and the letter from Krista. For a moment, he would not allow himself to read it, for he hadn't earned the right, he mumbled to himself. But even that resolve weakened, and for a few seconds, he was oblivious to his own misery, to say nothing of the sufferings of Branch, Bearchild, Fricker, and the Ranger with the bloody ankle.

His tranquility, though, was short-lived, for guilt soon roiled in his chest as he poured over Krista's declaration that he was a fine and noble man, a brave soldier. Within seconds, shame wormed its way into J.W.'s heart, and he snuck back into the barracks proper, pacing, struggling to conjure a strategy to undo the felony that had begun innocently, but quickly grown so critical.

He redressed in the clean uniform he'd worn to face Vogel, donned his second pair of boots, toes spit-shined to a fare thee well, and walked toward the door. He would do what the honor code at West Point demanded—report to the command shack and face his punishment. After a few steps, he came to the unopened mail scattered on the floor. On the top was a letter, the address scratched by a soul who suffered a dreadful tremor, or perhaps had barely learned the alphabet. Nonetheless, J.W. smiled as he made out the letters of the return address: Bessie Branch, 127th Street, Apt. 2H, New York City. J.W. thought for a moment about that neighborhood, summoning an image of the masses of dark faces fighting to survive one more day deep within the belly of Harlem.

He cracked aloud with a smile, "Branch has a mother?"

He gathered the envelopes and placed them in a neat pile on the bed closest to the door then walked out into the very early morning darkness, deep in thought. Nearing the mess hall, his trance was nettled by the thump of distant combat boots closing on Harmony Church. For a moment he was confused, not fathoming how the company could already be returning from whatever mistreatment the cadres had contrived for the second morning. Then he remembered that on some mornings, rumor had it, they would just run loops around Harmony Church for as long as it took to weed out a few more of the less sturdy. This might be a gift, he pondered, and his remorse faded as a plan to sneak back into Ranger School washed over him.

He stole through the icy fog, the black mist shrouding him as he loitered in the shadows behind the mess hall as the lead elements filtered into base camp. He was as still as a tree, his back pressed against the peeling mess hall. When the main body shuffled past, he waited twenty seconds then loped into the mass of those drifting on the edge of expulsion, imitating the beaten gait of the stragglers. Though expecting an iron hand to snatch at his shoulder and toss him aboard the meat wagon, nary a cadre nor Ranger detected the Rosie Ruiz of Ranger School.

The only crack in J.W.'s scheme came when the letter from Krista flipped out of his pocket. He stopped to pick it off the ground but, from the corner of his eye, saw Cowsen antagonizing the meandering few at the rear of the two-hundred-meter formation. He made a hasty decision not to retrieve it and turned to see Cowsen pick the letter off the ground, stare at it for a second, then stash it into his pocket.

Cowsen hounded the dwindling company for a few hundred meters then halted them in front of the mess hall, announcing that due to inflation in America, a horrible disease spread by greedy businessmen who had not been to Ranger School, the price of admission had risen to twenty pull-ups, and breakfast number two had been reduced to thirty seconds. Ranger Gillette, another small man, the one with thick, black-rimmed glasses attached to his head with an elastic band, raised his hand and was recognized by the first sergeant.

"Sergeant Cowsen, excuse me, but that calculates to an inflation rate of thirty-three percent on the pull-ups, and a similar percentage of deflation on food-consumption time. That's sort of like a crash on the stock market. Doesn't the military budget for the Ranger contingent?"

"Ranger Branch," Cowsen asked quietly, "what did he say?"

Branch looked up from his stupor and was about to ask the sergeant to repeat the question, but Cowsen barked, "All right, pull-ups and breakfast."

With the first rays of light, J.W. could see the uniforms of his compatriots were now thick with red mud from the single spin through the obstacle course he'd missed. He drifted away and found a puddle on the dark side of the mess hall, scooped out mud, rubbed it onto his bald scalp, and then into every crevice of his pristine uniform.

During breakfast, there was a commotion outside the mess hall, and the cadres were summoned to a formation of their own. It bought the cadets several minutes of peace, though they spent it looking at each other uneasily. When a handful of instructors returned, their faces were stone-like; not a word was spoken between them until one sputtered, "Fall in."

Twenty minutes later they sat in the bleachers. Half the seats that had been occupied the day before were now empty stretches of rotting planks. Branch and Weathersby, though they took seats far apart on a slat three rows from the top, were instantly discovered and dispatched back to the penthouse, now forced to sit backwards, legs dangling over the ravine. J.W. looked down for a trice and saw the soggy pair of cigarette packs maintaining their silent vigil on the valley floor.

Cowsen stood at the podium, silent for a long time. His face had become distorted., a hardness having transfigured the boyish, handsome face.

J.W. muttered, "Now what?"

"Rangers," the master sergeant declared flatly, "Martin Luther King is dead." There was a gasp. "Silence! He was shot by a madman, a bigot. Rangers, there is bigotry in America, and there is fire in America. Atlanta is burning. Washington is burning. But there is no bigotry in Ranger School, Rangers, and there will be no fire here, except the one that better be burning in your gut. You will drive on and become Rangers no matter what madness takes over America, no matter what price you have to pay. ON YOUR FEET! FALL IN!

The announcement saddened J.W, for he had grown up in New York City. A black woman had worked in his house as a maid, and for much of his childhood served as a surrogate mother. For years, J.W. had inferred that by watching his bed and meals being made by Posie, and by being yelled at by her when he picked on his little sister, he knew what it meant to be black. He looked with sympathy toward Branch but had to turn away quickly when the small man glared back at him with frozen, furious eyes. Branch finally spit violently on the red earth, the hocker landing near J.W.'s boots.

Cowsen spent some seconds surveying the faces of his few black Ranger cadets, concentrating on Branch and Kenyon. When they looked away, he quickly commanded the company to attention and then into a forced march. He screamed cadence for the entire pilgrimage, much of it in Branch's ear, but Branch had fixed his stare on a distant mound of earth. A single tear rolled off his cheek.

Cowsen eventually tired of tormenting Branch and trotted off to antagonize Kenyon. J.W. took the absence of Cowsen's authority to pull alongside Branch and console, "I'm sorry, man," but Branch glowered toward his hill as if he hadn't heard. J.W. slowed until he was abeam Kenyon. He mumbled, "I know you're really pissed off, but not all of us are like that. I am sorry."

Kenyon smiled weakly and nodded. "I know, man. Thanks."

At the edge of an empty, ice-covered expanse, they were given the order to halt under a mammoth tree from which dangled a wooden sign burned with the words, "Victory Lake." Behind the last rank of Rangers was another set of bleachers and a wooden shanty not unlike the command shack at Harmony Church, except here, wisps of wood smoke drifted from an exhaust pipe. Fricker looked out at the ice and groaned, "Christ, we double-timed all the way to the Atlantic. I don't know, we were goin' north, weren't we? Maybe it's the Great Fuckin' Lakes." At the water's edge stood a fifty-foot-telephone pole. Thirty feet into the iced-over ocean was another pole, both masts connected at their tops by an eight-inch-wide board that bowed at the center deeper than the rope over the gorge. Partway along the plank, sitting catty-wonkus, was a set of wooden steps, three up, three down, a warped, sideways Olympic victors' stand. Fricker pointed to a third pole, fifty feet farther into the water. He shivered, "Shit, that thing's a hundred feet high."

Cowsen cleared his throat. "Seventy-five feet. Rangers, you are at Victory Lake. You have made it this far—consider it a victory. You are the lucky ones. You will climb the first pole, start walking, and negotiate the horizontal plank, never touching it with your hands, NEVER touching it with your hands. Can you see the cable going from the second pole to the third, the one way out there in the lake, Ranger Fricker?"

"Yes, Sergeant."

"You will monkey climb that cable just like you did at the ravine. When you get to the little Ranger Tab hanging at the far end, you will slowly and carefully let your legs unwrap from the cable and ask permission to touch the tab, and then ask permission to drop into the lake. You will scream the word 'Ranger' as you fall the seventy-five feet."

Fricker raised his good arm. "What about the ice, Sergeant Cowsen?"

"What about it? Ranger Fricker, do not be concerned. As you know, we are interested primarily in your welfare."

Cowsen nodded to his confederates, and they about-faced to disappear into the hut. Within a millisecond of their exodus, an explosion of nuclear proportions shook southern Georgia. The shock wave slammed the company to the ground, into a crawling mass of mud-stained olive drab. A funnel of ice and a thousand gallons of frigid water lifted off the surface of Victory Lake. It arched back to Earth in a thirty-meter diameter blob. Every drop of it funneled onto the troops. After a minute, the men rose uneasily to their knees to shoo chunks of ice from their soaked, tattered uniforms. As the first man rose to his feet, Cowsen and his cadres emerged, dry, from the hut. He bellowed, "Ten hut. Fall in," then marched the company into a file at the base of the lakeside pole.

The first Rangers, the ten percent who forever reflexively pushed their way to the head of the line, climbed with lips turned in arrogance to the plank fifty feet above the surface of the lake. Lips puckered, though, as they reached eye-level with the narrow, horizontal board. At the tilted staircase, each man slowed and tiptoed up then down, arms swinging madly, seeking balance like a wounded bird falling from the sky. With the stairs behind them, they faced the second half of the warped plank, the end of which anchored the final cable.

That was where the monkey climb began. The sagging, ice-coated wire was no bigger around than a pinky finger. At first it was easy—downhill along the droop. The up-slope toward the orange and black Ranger Tab, though, would seize from them every gram of might that had not been beaten and starved out of them. Some of the initial warriors fell to the lake flat on their backs, the posture they'd held as their grip failed. Several of the gloating who made it to the insignia allowed their legs to undrape from the cable with a haughty snap. These must have been history and music majors who hadn't sweat through the basics of pendular physics, for as the weight of their legs swung down, so did they, hurtling sideways seventy-five feet into the already re-crystallizing water. The slam into the ice-thickened soup of Victory Lake moved even the staff sergeant, who, after cursing aloud each time a Ranger fell, rowed out in a dinghy and dragged the floundering troop ashore.

J.W. slid back a few spaces toward the rear of the line as attention fixed on Branch. The little man moved rapidly up the pole and nearly low-crawled as far as the steps, but touched one inadvertently. Cowsen roared so raucously, Branch lost his balance and fell sideways into the shallow water. He lay there for a moment, and one of the cadres cursed and started in after him, but Branch struggled out on his own. He hobbled back to the pole and reclaimed his position by shoving the next Ranger out of his way. He was back on the plank in seconds, blood dripping from his left ear. He ran up the stairs this time, stopping on the top step to spit into the water, the glob landing not far from the staff sergeant in the dinghy. He sprinted the rest of the way to the end of the pole, monkey climbed up to the tab, spit again, and announced, "Ranger Branch is dropping." Without waiting for permission, he let go and sailed gracefully toward the water, a sepia dart, leaving few ripples as he disappeared. It was so long before he surfaced, the staff sergeant expectorated into the water, shook his head, and piloted the rescue rowboat toward the last place the world had laid eyes on the haughty Ranger cadet. Branch, however, bubbled up, glared at the staff sergeant, and swam awkwardly to shore. He crawled to a seat in the bleachers, leaned back peacefully, and lit a Winston.

Each time J.W.'s turn neared, he waited for the Ranger on the board to begin teetering, drawing all eyes to the plank, then stole to the rear of the line. Cowsen, though, who had never taken his eyes off the steps, finally yelled over his shoulder, "Ranger Weatherman, get your sorry butt on that pole. You're next!"

A single nail fastened each rotting wooden rung into the pole, allowing the tread to rotate like a propeller. One had to grasp at the rung above with both hands and do a pull-up to keep from sliding back down the pole. Refusing to look down, J.W. fixed his eyes directly ahead, on the black droplets of congealed creosote clinging to the pole, imagining them to be the petrified bubbles of nervous sweat left by past generations of Rangers. The first few feet of the plank were easy, but the board narrowed quickly, as if an hourglass, and the miniature staircase vibrated in concord with the twitching of his legs. The steps were canted to the left, but more significantly, they had become glazed with ice. His hand brushed each step lightly.

Cowsen was silent until J.W. had traversed the steps and crept along the plank toward the far end, sweat pouring in the frigid air. As he was about to drop onto the cable and begin the scramble toward the Ranger Tab, Cowsen shouted to him, "Not so fast, Climb the Stairs Forever. Get your butt back here. Let's try it in reverse this time."

J.W. followed the order, conscious of the amplified shaking in his legs as he navigated the steps backwards. Again he touched them lightly for moral support, though Cowsen was silent, and J.W., assumed the first sergeant wasn't paying attention. He relaxed a tad, slunk to the end of the plank, reached for the cable, and swung out. As his quivering legs wrapped around the down-sloping wire, Cowsen hollered, "I didn't give you permission to mount the cable, Ranger. So get back down and try the whole cotton pickin' process over again."

This was the third time over the stairs, and he became bored with the exercise. He moved over them quickly, relatively upright, and was on the cable before Cowsen finished giving him permission to proceed.

J.W. was surprised at how much easier it was to hang onto a metal cable than the hemp rope over the gorge, but that was on the downswing of the parabola. At the low point, he turned his head toward Cowsen and smirked, but the man was busy taunting Kenyon.

On the up-swing to the Ranger tab, though, his arms began to burn as if sizzling bayonets were pressing against his triceps. He wriggled to lessen the pressure on his arms, and his fatigue pants became unbloused from his boots. The soaked hem quickly hardened in the winter gusts, leaving him insensate at one end and burning at the other.

J.W. considered his options. Clearly, he could go no farther on the wire, and he could certainly not drop into the lake from that height, but he could, he suddenly realized, crawl back down the cable, mount the plank, walk backwards across the steps, spit like Branch, trip down the rungs, and scream, "I QUIT THIS SHIT!"

He would lose only face, and that meant nothing. This wasn't China where all they talked about was losing face. And no one was paying attention anyway, each Ranger concerned only with his own longevity. J.W. was destined to wind up in Viet Nam no matter what he did, be it in eight weeks, eight days, or eight hours. It no longer mattered. Sooner or later, he would face the music for his foolish, impetuous decision to enter ROTC, and the compounding of that blunder by imploring the Pentagon to allow him to volunteer for duty in a war in which he had lost all interest. But, off he would go—anything was better than this. He would distinguish himself in combat and then laugh at the fools who survived Ranger School and wound up in Viet Nam anyway. He would earn his decorations in combat, not at the hands of a paper tiger corps of stateside cadres.

Seconds passed and his fingers blistered with the pressure of the line. His right hand loosened for a rest, and when he gripped more tightly with the left hand, his thumb was pierced by a metal spur poking out of the ancient cable. By reflex, his hand pulled away, and so did he, his legs forcibly wrenched from the wire as his torso swung down. J.W. Weathersby found himself hurtling toward the drink head first.

He opened his mouth to explain what had happened as the staff sergeant screamed up, "No one gave you permission to drop,

Ranger," but a blast of freezing water filled J.W.'s throat then drove into the pit of his stomach. His chest crushed as if a bear trap had clapped shut on it. The brutal pressure expelled, in a submerged belch of fraternity proportions, his last trace of air. It also squeezed an acidic squirt of gritty, intestinal juice into his mouth. Helplessly, he gulped another mouthful of water to dilute the putrid fluid.

As he threw his arms toward the surface in a last effort to clear a path to air, J.W.'s hand hit a piece of wayward ice, opening the lacerations on his raw knuckles. An ooze of dark purple leaked from the water-shriveled skin.

"Sergeant Hartack! Front and center. Talk to these sorry Rangers."

Hartack moored his boat and took the podium. Spit-shined, starched, and bone dry, he groused, "Rangers, you did not do well These exercises are for your benefit. But some of you did not fol- low our instructions. Some of you did not touch the Ranger Tab. Some of you touched the steps; some spent too much time under the water in an effort to avoid remedial push-ups. That is NOT a victory. You cannot expect to complete the training if you are not willing to play the game. And if even one of you decides not to play the game, like for instance, entering the water head first, you can't expect a ride back to Harmony Church."

All eyes locked on J.W. The staff sergeant nodded. "That's right, Rangers." He raised his arm and waved away a pod of chubby, dry GIs who had been lounging on a bench by the lakeshore. With the snap of Hartack's wrist, they flipped their half-smoked cigarettes into the lake, climbed into the cabs of their puke-green trucks, and fired the massive diesel engines to life. To a man, the Rangers inhaled deep clouds of the pearl-white, stinking exhaust, a pleas- ing bouquet, one that aroused dreams of lounging on the wooden truck beds for the multi-mile journey back to Harmony Church. Some of the Rangers who had been asleep during Hartack's cri- tique drifted unconsciously toward the vehicles, but they were rousted from their insensibility by the staff sergeant's roar. "HALT!

I told you, you didn't earn the right to sit on your lazy, maggot asses. Rangers, fall in."

As the troops coalesced into a shivering, rectangular formation, the trucks pulled out with empty cargo beds, save for the morning's washouts. The rigs left behind a huge cock's comb of red dust that hung in the glacial air. Cowsen waited for the cloud to waft over and stick to the dripping faces and sodden fatigues of those who had endured. He took his place in front of the column. As the dust consolidated on the uniforms, he laughed, "You Rangers look like fried chicken. On the double time, forward march."

It took less time to move through the mess line that evening. There were many fewer Rangers, 68-B down to eighty- two souls. It would have been eighty-three, but for Kenyon, who had told his Ranger buddy just that morning he'd never make it, "'Cause I'm black, and there sure as shit *is* prejudice in the army. No black-as-coal mess hall cook gonna ever wear the Ranger Tab."

According to rumor, the cadres had begun tormenting Kenyon on the obstacle course that morning. Maybe it was because the man was huge, and the cadres loved to pick on the big guys, for as a class of men, the instructors suffered desperately from a deep-seated illness—SSS, the Small Shit Syndrome. Kenyon also looked strong, like a bear, like Sonny Liston, and the cadres must have thought because he was black, enlisted, and big, he was in shape.

But Kenyon was there by only dint of his heart, his refusal to accept the, "No fuckin' way," with which he was answered each time he applied to enter The School for real. Kenyon was but a cook at Harmony Church, and though the staff, kitchen workers, cooks, drivers, and clerks assigned to Ranger Training Battalion wore black berets, it was without the Ranger tab. As enlisted men, there was little chance they would be chosen to enter the actual training, though in the annals of the school, it had come to pass, although rarely. Those of the common workers who'd graduated had been elevated to Ranger legends, their names spoken in reverence by the older cadres, examples for the present sorry lot of cadets.

Perhaps Kenyon had dreamed of wearing the real beret someday. Doubtless, that is what he dwelled upon while slapping loose, lumpy mashed potatoes onto the mess kits of the chosen for those two years of his indentured service. Maybe he even dreamed of finishing Ranger School, muddling through Viet Nam, and returning to the ghetto wearing the Tab. He would stand forever above the rest. He'd avoid the depression of the unemployed, those burning away their time guzzling cheap talking wine, rapping in front of some decaying bar in West Philly.

Conceivably, Kenyon, if he graduated from the school, would escape the entrails of Philadelphia all together and become somebody. That, of course, could have only been the case if he had lived beyond the trek from Victory Lake back to Harmony Church. Kenyon, however, after a few miles, began to walk a serpentine course, drifting from the main column, tripping on pebbles, dazed, babbling to the Rangers who nervously defied the staff sergeant and tried to carry him.

Yet Staff Sergeant Hartack wouldn't give up, as if the saga of Ranger Smith had been for naught. He poked at Kenyon with a stick, screaming, "Move your ass, fat Ranger. Drive on." Hartack began to froth at the mouth. "Tax payers ain't gonna piss away money to carry a loser like you. You think you get a free ride 'cause they assassinated Martin Luther King? Look at your sorry ass, spook. You ain't even sweatin'."

When Kenyon collapsed, Hartack kicked him in the belly. And when Kenyon closed his eyes, the staff sergeant kicked him harder and then harder, shrieking, "The taxpayers ain't about to pay no twenty-five thousand dollars to send no brownie to Ranger School. Git off your fat fuckin' ass."

According to the stragglers, Kenyon crawled away from Hartack, finally rolling into a drainage ditch. A moment later, all movement ceased.

"You can't die from getting kicked, can you?" J.W. questioned Fricker incredulously during the afternoon forced march.

Fricker waited until the double-time was over to gasp, "I heard he was killed right there on the side of the road." He paused to catch his breath then reflected, "Kenyon's lucky. He's out of the crap. At least he's resting."

"That's illogical." Gillette, the little man with the thick, horn-rimmed glasses, interjected. "You think you need to die to get out of Ranger School? All you have to do is say, 'I quit.' You don't have to be so melodramatic."

Fricker considered Gillette's words then muttered, "All I know in the world is that now there's near a hundred lucky guys, and a handful of chumps with fifty-four-and-a-half days and a wake-up to go. You tell me who's better off."

Staff Sergeant Hartack wasn't at the pull-up bar that night. Cowsen took his stead but was unusually subdued in demanding the Rangers execute twenty-six pull-ups before chow. The staff sergeant wasn't in the mess hall either, and the Rangers nodded knowingly, albeit surreptitiously, comfortable the cadres had suffered a powerful lesson, one that would insure the remaining cadets a humane passage. Class 68-B would be carried on the coattails of Kenyon's agony.

Though the spuds were wetter than usual, and the cooks slammed them onto the trays so hard much of the meal splattered onto the uniforms of the silent Rangers, bitching about the rations was hushed. J.W. guzzled milk, but it was warm and even more bitter than it had been that morning.

Evening formation commenced with pull-ups—one man quit at number eighteen. A solemn Cowsen then marched the company south toward the bleachers for a night navigation problem. As they broke into the double-time, Branch's face twisted and his lips closed even tighter than when he had heard about Martin Luther King. He grumbled to himself for a mile or two, speeding up toward Cowsen, then dropping back, muttering more and more stridently to himself. He finally broke ranks and charged to Cowsen's side.

"First Sergeant," Branch thundered breathlessly, "Kenyon was lynched 'cause he was black. What are you going to do about that, First Sergeant?"

Cowsen responded by accelerating the pace, the rhythm of his barking slowly eroding from a cadence into a monody. When he ran faster, as if trying to escape, Branch pursued and came within inches of the man, but Cowsen's stare remained fixed into the glacial darkness. For another mile he tolerated Branch's shadow until the Ranger cadet again demanded, "What are you going to do about it, First Sergeant? Wait until the white man kills all of us off?" Cowsen spun his head to the right and stopped short.

As the company passed the two black men, Cowsen howled, "I told you. There ain't no bigotry in Ranger School. Now either get your sorry butt back into formation, or get out of this school. You think you're gettin' special treatment 'cause you're a black baby?"

Cowsen pivoted into the night, running, not with his troops, but from them, disappearing into the gloom. Branch remained by the side of the road, glaring contemptuously as the leaderless company of cadets passed in perfect obedience. He finally loped slowly toward the column, but suddenly spun and chased after Cowsen into the night.

Captain Vock met the company at the bleachers. He stood nearly at attention, silently waiting as deep, grumbling, angry voices approached the bleacher clearing. J.W. turned his ear toward the sounds, and as he fought to hear, his eyes caught the moon glint of cigarette-pack plastic at the bottom of the ravine.

When the voices suddenly stopped, Branch appeared from beyond the shadow of the bleachers. As he climbed to the top row, Cowsen and the staff sergeant emerged together, though their heads were turned slightly away from each other. Cowsen stepped to his podium. Vock slid back into the night.

"Rangers, welcome to the map course. You and your buddy will be issued a set of compass directions and distances to various checkpoints. At each checkpoint there is a prize—instructions to the next checkpoint. At the last station you will find directions back to Harmony Church. The faster you complete the course, the sooner you will be in your rack. Questions?"

J.W. raised his hand and asked, "Is this some sort of scavenger hunt, like when we were kids?"

"I don't know. I was never a kid, but I'll scavenger your butt, Ranger, if you don't close your wise-mouth behind. Now, FALL IN, all of you."

The staff sergeant swaggered up to J.W. "You tag with Branch and his buddy until someone else drops out or dies."

Branch heard the order, muttered to himself, then walked to Bearchild, whispered in his ear, and the two spun toward the tree-line. Bearchild sniffed the air, took a compass reading, and melted silently into the forest. Branch hurriedly ducked in behind him.

Hartack carped at J.W., "You're already lost, Ranger. Get your ass in gear."

J.W. kept up with Branch and Bearchild for a couple of miles, but his eyes slowly became choked with what he believed were giant cobwebs. He tried to brush them away, though his eyes refused to stay open. Each time Bearchild stopped to take a compass reading, J.W.'s head dropped and, while still upright, he drifted off into an uneasy sleep.

It was hard to know if he was awake or just dreaming when he heard Branch ask Bearchild, "Hey, man, where'd you learn to track like that?"

"On the res."

Those were the first words J.W. had heard, and amongst the only he ever would, from Bearchild, one of the few Native American officers in the United States Army. In fact, as J.W. thought about it, though there had been a number of American Indian enlisted men, all Viet Nam returnees, who had been shoved into the 6th Cav to await discharge, he realized he had never heard of a Native American who'd received a commission. He peered at Bearchild in the meager moonlight and saw the resemblance to the Native American soldiers who had served in his platoon at Fort Meade. J.W. remembered that the Caucasian troops used to laugh and call the Indian soldiers "Chief," but looking into this man's piercing, black eyes, he could not imagine anyone uttering those words to him.

The three walked on, Bearchild staring at the ground, occasionally studying the trees, picking off twigs, tasting them, sniffing the air, and from time to time, grunting quietly when he was pleased with what he had unearthed. Branch, however, jabbered endlessly as the fatigue pressed in, first bitching about the cold, then, with a few hills behind them, about the heat. After two hours of listening to Bearchild smelling the air and Branch grousing, J.W. finally asked Bearchild, "What the hell are you sniffing?"

"Smellin' for urine."

Branch mocked, "Yeah, like you can tell the difference between animal and human piss. Right! You're just wasting calories, my man. Let's just get this shit over with and catch some Zs. I ain't here for no nature walk, Bro."

"Man, I wanna head back, too," J.W. moaned. "Maybe smell for the latrines at Harmony Church. How 'bout that? Find us a way to the barracks right now. We'll tell 'em we ran the course in record time, or some such bullshit. I don't care anymore. I wanna go home."

Bearchild nodded, took a compass reading, and set off on a straight northerly tack back toward Harmony Church. For twenty minutes the three busted brush like men possessed, but when they encountered a thick patch of razor brambles that stopped them dead, J.W. was astounded to see these two tough men of color no more able to move through the barrier than he. They also struggled with the prickers, the hunger, and the pressing fatigue; blood from the growing number of gashes on their arms and faces was just as red as his, glistening in the bare moonlight.

It suddenly dawned upon him that though they were children of disparate pasts, that night they would not be afforded an option— they would bleed and freeze together, and their legs would burn with the same pain under the same sinister sky. For however long that night was to last, all three were to be at the mercy of the same grand spirit that had denied them a simple fate.

For an instant, J.W. smiled, grasping the notion that the brainless army had created a trivial miracle, the no-nonsense melding of three incongruent civilizations. It was a deed seldom encountered in the real world. He beamed with pride, with the same contentment he had enjoyed so long ago at West Point in the men's room, five days before.

After another hundred meters of tearing through the blackberry thicket, Bearchild declared, "Acacia."

He turned back and took a compass reading that put them on the hypotenuse to meet the prescribed course. They had lost an hour, though with the hope that they could cross more easily through the Acacia forest, the team moved with a renewed spirit until they ran into an even more deeply wooded area, one without thorn bushes, but laced with hoots, screeches, and hisses. J.W. muttered, "I don't like this shit at all."

Branch stammered, "Calm down, Bronx boy. That shit's bein' piped in by the cadres, like forest Musac. Those fuckers never miss a chance." But what J.W. couldn't miss was the waver in Branch's voice.

The shrieks, growls, and yelps even caught Bearchild's attention. In the mist-filtered moonlight, J.W. saw Bearchild's pupils dilate cavernously as he warned, "Stay awake."

At midnight, J.W. celebrated the beginning of a new day by pausing to take a leak against a huge, old-growth oak, keeping the crunch of his compatriots' footsteps in the crosshairs of his consciousness. The next recollection, however, was not of their boots, but of a moist, warm feeling on his leg, and then the piercing lack of human-generated sound. As the moon became engulfed by a black cloud, darkness the likes of which he had never dreamed descended upon the forest. Only his ears saw. J.W. sought to find Branch and Bearchild by targeting snapping twigs in the distance, but after a few minutes, even that thread of reassurance unraveled. He realized the cracking was coming from all directions, and that the animal din was closing in.

He sat for a moment on a rotting log recalling the myriad speeches he had slept through in those first, painful hours at Harmony Church. He knew Cowsen wouldn't have sent eighty-some novices into the mouth of the lion without the wherewithal to survive, but all he could remember was the primary rule for that night's exercise: "Rangers, you do not speak on a mission. That means absolute silence unless you are spoken to by one of the cadres."

"Screw 'em," J.W. hissed. He took a deep breath and shouted in a whisper, "Hey, where the hell are you guys?" The deadfall rustled, and he moved forward quietly, relieved to hear Bearchild's heavy breathing delivering another urologic diagnosis. But when J.W. stumbled into a tiny clearing, he caught the sudden flash of luminous eyes the size of silver dollars reflected by the last sliver of moonlight. In the utter darkness, he had time to ponder the fact that neither Branch nor Bearchild had luminescent eyes. Nonetheless, he moved forward and cursed, "Will you guys slow down, goddamnit." He waited, hands pushing his ears forward to listen for the cluck of Branch's tongue.

When the moon appeared again, J.W. saw more of the head that held the glowing eyes, and as the clouds parted further, the entire body of a great buck deer became visible. Catching J.W.'s scent, the beast's nostrils flared, and it moved slowly backward to find itself cornered in a crevice of the forest. Eyes locked on J.W.'s motionless frame, it shivered violently, then backed another step into the timber, its hindquarters whacking a spiked trunk of dead-fall. The monster shuddered and reared on his hind legs, slicing at the freezing air with razor hooves. When it shot frontward, head down, the rack aimed at J.W.'s chest, Weathersby dropped to the forest floor, clawing as he had in the mud-filled tube at the obstacle course. This time, however, his fingers dug fiercely into the humus, and he found himself propelled to a defensive position behind another great oak. The buck followed, banging at the massive wood trunk with its antlers. J.W. spoke aloud tremulously, "Shit, the thing can see at night."

J.W. remained perfectly still as the animal's thick, musty scent burned in his nostrils. Despite his enervation and fear, it struck J.W. that the buck could not trample, or even gore him, if he just kept the massive oak between them. With that breathing space, a sense of exhilaration built inside his chest. It was, he supposed, what soldiers returning from Viet Nam had tried to tell him they missed stateside. So this is what it was like to face a formidable enemy, one strong enough to take his life with a single blow. In his childhood, he would have run from a threat so powerful, but he realized there could be no retreat that night. He would fight or die.

His first thought was the M-14, but the cadres had not seen fit to issue live ammunition, aware the majority of the class would be decimated in minutes by friendly fire had they been rationed even a single round for every fifth man. But it struck him that if he could remain calm and patient, and aim fearlessly at the last second, he could stop a bear with what he held in his hand. So, he gripped the rifle by the barrel and raised it exceedingly slowly until the beefy, wooden stock was nearly twelve feet off the ground. Nothing, he snarled, could survive a blow of such violence. His arms tensed to the rigidity of steel as he readied to shatter his enemy's skull.

"Okay, goddamnit. You wanna fight, mother fucker?" He peeked around the tree. The buck was so close, it was peering at him with one eye. J.W. shuffled slowly along the peat of the forest floor, locked his torso, and took a final breath before the downstroke. His heart, though, clutched at the last millisecond as he realized that, with the animal's head canted, his blow would be blocked by the endless horns. The strike would have to be across the forehead, but as J.W. re-aimed, the buck turned and wandered slowly into the blackness.

J.W. waited until he could no longer see, hear, or smell the animal, then leaned breathlessly against the oak and pulled a pack of cigarettes from his first aid pouch. The instant the match flared, however, the stag reappeared, leapt into the air, and slammed its rack into the tree. A spike of the creature's antler speared the moss-covered trunk and fractured with a crack so piercing, J.W. thought a rifle had been fired. The animal kicked and screamed then backed off, and J.W. waited for it to drop from the bullet, but it charged a second time, and J.W. was barely able to scramble behind the tree. As the maddened beast's head hit the trunk again, it became unsteady and dropped to the ground, emitting a high-pitched bleat that pierced the night.

J.W. sprinted through the forest, deeper and deeper, ears cocked for any living form of life, even a Ranger. He heard the groan of a distant vehicle and headed west toward it, but when the roar now came again from the east, he realized he'd been enticed not by a car, but by the rumbling of thunder. He slowed his retreat, sat, smoked his last cigarette, then let his eyes close, promising himself it would be only for a few seconds. He awoke to the sting of a buzzing, crusted insect on his arm and took to his feet, walking for, he was sure, hours, backtracking, forward tracking, pacing for a time in circles, until happening upon a macadam highway. The edge of the road served as a guide as he marched away from the lightning-charged eastern sky.

He went on for another hour, reviewing his life, musing morosely of friends in graduate school and those working for salaries, buddies building families, safe, warm, unafraid of their tomorrows, or of their next months. He thought sadly about the years of study he had put into his engineering degree, a sheet of paper that would be worthless in five years when he was released from the army. He was learning to do push-ups and stare down wild creatures in the wilderness. He feared that would be all he'd know when he was discharged, if he lived that long.

He cursed himself for being drawn into the army's web of promises at Sterling College. He thought back to the day in the early 60s when the ROTC commanding officer addressed the freshman class in Warner Building 121. The man had painted images of the Berlin Wall and the Cuban Missile Crisis to keep his teenaged audience awake. The gray-haired colonel intoned, "Gentlemen, America, the country that has given us so much, for which we will fight and die, is in trouble. There is a ground war brewing in Asia." His facial muscles tightened. J.W. was sure the man was staring at him. "Mark my words. It is inevitable. Everyone is going. Do any of you want to go as privates toting an M-l?" The colonel surveyed the room quickly and continued before any of the children in his audience had a chance to consider his query. "Hell no, you don't! I'm here to tell you, you want to go as officers."

J.W.'s hand shot up. "Sir, how do I get to go as a pilot?"

"Good question, very good. Meet me after class, and I'll tell you, son," the man offered warmly. And within an hour, J.W. had signed away nine years of his life—three more in ROTC, a year in flight school, two as a ground-pounder lieutenant, and three more tacked on for the privilege of having been selected a Distinguished Military Graduate.

While the army had kept the colonel's covenant until J.W. graduated from flight school, he hadn't seen the inside of a cockpit since. There had been armored cavalry units, riot control units bridling hippies at the Pentagon, airborne units shuffling through Fort Bragg before dawn, and now Ranger School, each assignment further and further from the year of work he had burned away learning to fly, each year and each task since flight school just another notch on the long list of disenchantments.

What he was sure were hours and miles passed on the highway, each laced with loud curses for having relinquished his best years, for not having quit earlier that day. "There's nothing keeping you here," he yelled louder with each hill. "I've seen dozens of good men, a lot better than me, get kicked out over the past two days. Not a single one of them's been beaten. There ain't no record in heaven of what happens in this bullshit school. And even if there is, I got a whole career to earn redemption. Man, I quit this shit."

When lights flashed in the distance, he started toward them, but they soon disappeared. He sat on a big rock by the highway, which had become wider, now with a white line down the center. He uttered a firm oath to resign the moment he set foot back in Harmony Church. He reached for a cigarette, having forgotten his pack was empty. He had wanted to rest and smoke and think about his future, but his mind was clouded in such exhaustion, he saw nothing but gloom. His head dropped. In seconds, he saw nothing at all.

"Hey, soldier, you okay?"

He heard the sounds but thought it a dream and rolled onto his other side. The voice boomed again, and J.W. opened his eyes to the glare of stationary headlights.

The voice interrogated, "I said, are you okay?"

J.W. shook his head into consciousness, answering, "Yes, sir. I'm okay. Just need directions back to Harmony Church. I must have dozed off."

"Get into the car. I'll give you a lift before you get run over. You're laying in the middle of the road."

"I can't do that, sir."

"You have to get out of the road, soldier."

"No, I mean I can't get in your car."

"Why not?"

"It's cheating to accept a ride."

J.W. rose to his feet stiffly and stumbled along the highway, though the civilian drove in reverse alongside him, and after a few yards spoke kindly, "What are you talking about? Hop in, son, before you become a casualty."

The timbre of the words, though spoken soberly, were more a command than a choice. Reflexively, J.W. opened the door.

"What the hell. I'm done with the training."

The man was dressed in jeans and a logger's jacket. He turned the car around. "At least you were headed in the right direction." The man's nose wrinkled, and he cracked his window. "Hey, are you one of those," he paused and spoke with admiration, "Rangers? That why you're trying to get back to Harmony Church?"

"Yeah, I'm a Ranger."

The man hesitated for a moment and cleared his throat. "Why are you so far from your unit this close to dawn? I thought you guys were supposed to have buddies. Don't tell me we left him back there."

"My Ranger buddy? No, he's dead, I think."

"Dead?"

J.W. patiently recounted the field compass problem, and how he'd become separated from his temporary buddies, and then he spoke of Smith and Kenyon, and how he would resign in another hour and be on his way to Viet Nam. J.W. glanced sideways, expecting the man's eyes to bulge, his jaw to slacken, though he remained expressionless as he asked, "How long has this hell been going on?"

J.W. thought for a moment. "I'm working on finishing day two, sir."

"Whew! Sounds awful." He was silent for a while then asked, "You mean to tell me a man has to work that hard to get a Ranger Tab? Why? Army doesn't give a damn about you, does it?"

"Look, sir, I'm tired, and I'm hungry. Do you mind if I just rest for a little while?"

"Sounds like you've been through a lot. You made it past the cuts. You've proven yourself. It's time to go into combat and show 'em what you've learned in forty-eight hours, huh? Sure, take a rest."

J.W., hypnotized by the colors of the Oldsmobile Ninety-Eight logo on the glove box, nodded off in the warmth of the leather seats and the mellow, green glow of the dash lights. "Say, Ranger, figure your men are going to get tired and pissed off in Viet Nam in a few days?"

J.W. sat up. "Yes, sir. I'm sure they will, but that's combat. Viet Nam means something. It counts."

J.W. lowered his head and soon dropped off, tranquil for the rest of the trip. He was shaken awake at the base of a hill near Harmony Church. As he crept out of the car, the civilian pointed the way back to the command shack. "Up there, soldier. I think that's the Ranger School. No one'll see ya. You'll be able to slip in and resign, no sweat. And I don't blame you. I remember the war. How it was in Europe. They didn't care. I can't figure why the army does these stupid things. What is wrong with them?"

J.W. was about to agree, though felt uneasy, even piqued by the man's words, and took a step toward the command shack. By the time J.W. turned back to acknowledge the man's kindness for hauling him off the centerline of a Georgia highway, the power window had rolled shut, and the car was off toward main post.

As J.W. crested the hillock, he saw Cowsen lounging in a lawn chair by the command shack, peacefully whittling a green stick. Staff Sergeant Hartack was on his back, snoring, on a nearby bench. J.W. approached Sergeant Cowsen to demand an audience with Captain Vock. Legs wobbling, he stood above the first sergeant and took a deep breath to utter his final words of Ranger School.

Cowsen sat up and stared at J.W. He shook his head in disbelief. "You the first back? Good Lord. Where's your Ranger buddies, Bad Weather? You screw 'em up, leave 'em out there to die or somethin'?"

J.W. took a deep breath to answer, but the squish of muddy boots and bitching intruded as four Rangers wandered up the hill toward Cowsen. The first sergeant rose, pushed his chaise lounge into the command shack, and grabbed his clipboard. The men lifted themselves into attention to report in a jumble of names. "Start again, one at a time. Give me your name and starting time, and stay in order of when you got up here." Cowsen paused. "Weatherman, is it really you? First pigeon to fly home?"

"Yes, First Sergeant, and it's Weathersbeee."

"Weathervane, just mind your mouth and move your butt into formation. Go on."

J.W. walked off cursing under his breath for having muffed another opportunity to escape. He growled, "Mother fucker."

One of the men leaned toward J.W. "How'd you get up here so fast, man? We skipped the last two stations, and you still beat us. You got smokes?"

"Nope, didn't bring any. Knew I was going to run the course. You got one?"

J.W. took a seat on the frozen ground. Not spotting a cadre, he leaned against the command shack. "If I do it now, the bastards'll make me wait to get breakfast until all the real guys eat. I'll do it after a few slices of bacon."

He walked back to the first sergeant. "My Ranger buddies are fine, it just so happens. Both of 'em are on the way. I got a little excited and ran the last mile or two. I couldn't wait to see you again."

Cowsen's lips twisted skeptically, but with teams of bedraggled Rangers now floundering up the hill, the first sergeant went to work crossing names off his list, dozens and dozens of them. Two minutes later, the rest of the men surrounded Cowsen, pushing and shouting names. J.W. was astonished that the bulk of Class 68-B had converged on the first sergeant within five minutes of each other.

The inventory of his pupils nearly complete, Cowsen stared at the clipboard, glanced down the hill, then turned to the staff sergeant. "Hartack, go out and round up the ten-percent." Cowsen walked to the door of the command shack, held up the stick he had been whittling in the moonlight, stretched with great theater, and screamed, "Fall in! Rangers, you are going to be split into official platoons."

A mob of cadres descended out of the night to dash into the mass of muddled commandoes, jostling them into squads and fire teams. J.W. was shunted into a slot flanking Branch and Bearchild. Branch's face twisted as if he had just been told to prepare for a root canal. He whispered, "I thought you was dead, honkey."

J.W. turned away.

Cowsen announced, "Rangers, your vacation is about to come to an abrupt end. We're gettin' close to the core."

"The elite," Fricker whispered, rolling his shoulders.

"Let's see how elite you feel in a few hours," Cowsen hissed. "The rigorous training is about to commence. Gentlemen, we are about to set off on our first challenge of Ranger School, a tactical problem, a real combat mission. This is why y'all volunteered for this training, to learn how to fight. Your enemy, or aggressors as we will call them, are Green Berets. They're the army's best, because in order to become a Green Beret, you first need to graduate from Ranger School. We've lost a lot of men so far. There'll be more. Drive on, Rangers. It's all up to you. See you at the bleachers."

As the company was commanded to march at the double-time, J.W. groused a decibel or two louder than he had intended, "What about breakfast, goddamnit?"

Cowsen screeched the formation to an abrupt halt. He charged into the formation, stepping over the squirming troops who had been nearly asleep during the run and had collapsed into the men to their front. He prickled up to J.W. "You left your Ranger buddies out in the jungle to die. You abandoned them. You do not deserve to eat. And because you don't deserve to eat, your company does not deserve to eat."

As the unit restarted the run, J.W. swore to himself that Cowsen's most recent humiliation was to be the last straw. He would utter the magic words, right there, in the middle of the run. "That's it," he howled, head bobbing madly. "Fuck 'em. I will not tolerate another moment of this shit." He'd uttered that oath to the stranger in the Oldsmobile, and so would it be executed.

"Now!" he belted aloud, at the top of what was left of his voice. A few of the bedraggled looked about, but only momentarily. Weathersby forced his legs to churn faster, pursuing Cowsen, but they came to a hill and the first sergeant sprinted off alone toward the crest. When J.W. accepted he could not catch him, he let his legs slacken and his body crumble to the dirt road. Several Rangers hurdled his prostrate form; others, who had been running with eyes closed, tripped over him. Though they cursed, no one spent the calories to look back. J.W. didn't care. Ranger School was done. He had wrested control of his destiny.

At the bleachers, J.W. took his balcony position on plank Number 10, but this time, he sat with a back as straight as a plebe eating square at West Point. His mind boiled until the rebellion spread to his legs. J.W. Weathersby sprang to his feet and started down toward Cowsen, propelled by a need to stare down his parents, his neighbors, his teachers, and coaches, the spiteful to whom he had relinquished his childhood. Their malevolence was now distilled in Cowsen's arrogant face, and J.W. was going to confront those demons by defying Ranger School, the goddamn army, and the world, mano a mano.

On the way down the boards, his mind spun in preparation for the first brazen act of his life. It was a rite of passage he could no longer postpone. From that moment forward, he would never again play the weak-willed crybaby. At plank Number 8, he belched, "Them days is over, goddamnit."

At plank Number 5, though, he sensed his body beginning to shrink, his despondency rising to crush the daring. He slowed, and as he reached plank Number 3. His mind shot forward to Viet Nam, the old craving to rescue his country from the menace of communism bolting from his heart in a gush. It was not ugly scenes of war that confused and frightened him. It was the hopelessness of accepting that he was only fooling himself. He'd made the pledge to change a million times in the past, yet nothing had improved. He knew that when he got off the plane in Saigon, only J.W. Weathersby would be there to greet him, and even in combat, it would be same old story of frailty. He would remain a meaningless device, second tier, controlled forever by those with the right stuff. He stopped for a moment at the last step, demanding a return to the bravado of the tenth plank, the courage to move forward and cut Cowsen and all he stood for with a dagger. As his first foot hit the icy soil, though, the resolve again drowned in the futility of ever changing.

He turned to look back at the Rangers, most of whom were exploiting the interlude to close their eyes and let their heads bob. These were the warriors the man in the Oldsmobile believed had been chosen to lead America's sons into combat. J.W. pondered that, a moment ago, he could have been one of them, for it appeared to all the world, save Cowsen and J.W. Weathersby, that he had proven himself and made the cut. But the truth roiled in his heart, and he knew he had only obliquely earned the right to have starved for three days. It was time to go to Viet Nam, and so be it. That, too, would pass. He wanted no more of the army, of its advanced training and career guidance. He was not destined to be a military man. He would become something else.

When he turned back toward Cowsen, the first sergeant bristled into rigid attention, though his face was without expression.

There was no malice, no grace, only carved stone. Perhaps a cloud parted, for a fragment of J.W.'s darkness lifted, and J.W. saw the image of his wife, and then heard the timbre of her voice. She had called him a noble man and a brave soldier. It caused him to waver, and in that instant, he sensed an invisible hand on his shoulder, a new force, silent though compelling, a fist stronger than his despair. Its power turned him, and he found himself about to remount the first plank, but instead of climbing back up to the balcony, he settled in the front row between a cadre and a Ranger, both of-whom slid apart perfunctorily to allow him a seat. Though his spirit was as dark as the winter sky, and the temperature had not lifted above the low 30s, the heat that radiated from his trembling limbs was a luxury. He waited for the order to return to the high seats, resolving to swagger back up when it came, but Cowsen resumed his lecture and did not look at him then or over the next hour.

When the Class 68-B moved out after dawn, it was in formal platoons, each with a separate mission. Branch, Bearchild, Fricker, Weathersby, and Morelotto, the Second Fire Team of the Second Squad, and the other shaven heads of the Second Platoon, were issued orders to attack aggressors entrenched on the banks of the Upatoi River, miles from the bleachers of Harmony Church.

"Expect heavy resistance from the enemy," Cowsen warned. "If captured, you will provide nothing other than your name, your rank—of which you Rangers have none—your serial number, and your date of birth." Cowsen went on quite earnestly, "The security of the nation is riding on your shoulders. This is a serious mission. Intelligence reports indicate actual enemy troops, communists, lurking in the woods. They are preparing to take advantage of the chaos a goin' on in America. You are to be the first soldiers to challenge them."

Gillette spoke to no one in particular. "Must *be* anarchy if they're counting on *us.*"

Cowsen sauntered up to Fricker and tapped him on the shoulder with his whittled stick. "You are the first victim of the course, Ranger, the first student platoon leader. Plan an attack."

Cowsen handed him a manila envelope stuffed with military papers stamped TOP SECRET. Fricker shuffled through the stack, put the loose sheets on a big rock, then stared uncertainly at Cowsen. "Hey, wait a minute. That's twenty-five kilometers from here!"

Cowsen glared but took a deep breath and turned away. At that moment, a dawn breeze sent Fricker chasing the classified documents over and under the bleachers. A few sheets were sucked by a vortex into the gulch. The first sergeant, who had not turned back to look at Fricker, leisurely picked his teeth with a weed. He called over his shoulder, "Better get your butt on a stick, Ranger. The more ground you cover during daylight, less chance you got of gettin' late and lost. You got three sees to do before sundown."

Fricker stopped in his tracks and called to Cowsen casually, "Okay, Sarge, I give up, what's a see?"

Cowsen walked farther away, so Fricker went to the staff sergeant and asked, but Hartack snapped, "Ask First Sergeant Cowsen, Ranger. He's the boss. He's the one that knows everything."

So Fricker went to J.W. and asked, "Hey, man, what's a see?"

"I don't know. Maybe they can't spell, and it means kilometers, like Ks."

Branch pretended not to listen but couldn't contain himself and sprang to his feet. He snarled at Fricker. "A guy climbs a tree, picks a point as far as he can see, and that's one see, man. Don't you people know nothin? Now move your ass. I'm hungry." Fricker shoved the salvaged papers into his rucksack, raised a fist above his head, and screamed, "Follow me." They headed for the wood line. Second Platoon had once again begun Ranger School.

For hours they followed dutifully behind Fricker, busting and hacking through the brambles and thickets of southern Georgia, an army meandering without a battle plan, thirty of the country's elite searching for the Upatoi River, a geographic aspect that Morelotto sneered several times existed only in the minds of the sadists running the school. At the edge of another great forest, Fricker scratched his head and asked J.W. for a position report.

"Fricker, to be honest, man, I'm not much use when it comes to a map and compass. I'm left handed. We can't spell or tell directions."

"Then how come you were the first one back from the compass course last night? Look, if you don't want to help, just say so."

"No, man, it's not that. See, last night, I ahhh..."

"Well, then, gimme a hand."

Fricker shoved the map into J.W.'s face. Weathersby's index finger wavered for a moment then settled on an open area. "But it's a WAG, man, a wild-ass guess." After a quick peek, J.W. babbled, "On the other hand, it looks big enough to bivouac and chow down. That's the most important mission you got, Fricker. You gotta care for your men. We need to eat. Army travels on its stomach. Tell Sergeant Cowsen that's where we are, and that you want to feed us."

Fricker nodded and searched furiously for the orders specifying a rendezvous point with the mess personnel. "Look harder, man," J.W. beseeched.

Cowsen drifted over and opined that Fricker must have lost that information to the wind. "Show me on the map where we are, Ranger." But Fricker had forgotten where J.W. had pointed, so he closed his eyes and pointed to a dot on Cowsen's map. "Ranger, are you sure that's where we are?"

"Yeah, that's the place, Sarge."

"Ranger, I'm going to cut the tongue out of your cotton pickin' head if you call me Sarge or use the word yeah one more time. You got that, Ranger? Now get your slim butt a goin'."

"Yeah, ah, yes, but what about dinner?"

"Yes, what?" Cowsen boiled.

"Yes, Sergeant. Can I ask about dinner?"

Cowsen relaxed and answered softly, "Yes, Ranger, you may ask." Before Fricker could repose the question, Cowsen walked off to the back of the column.

Fricker stared at the map, finally raised a fist, and again bellowed, "Follow me."

They ducked back into the tree line and trod for hours aimlessly. Every few minutes another troop whined, "When we gonna to stop and eat? It's been more than twelve hours."

After miles of harassment, Fricker began countering each entreaty with, "I don't wanna hear it. Just drive on, Ranger."

As darkness closed in on Second Platoon, Cowsen touched Fricker on the shoulder with the stick. "You're dead, Ranger. Snake bit." Cowsen spun about, bent forward, and swiped Ranger LaVoy in the head with his now whittled-thin bough. It wasn't because LaVoy was excessively short that Cowsen bowed to beknight the next student platoon leader, it was because LaVoy had exploited the respite to drop to the wet earth and sleep. Unaware that the green-stick slap signified elevation to a position of leadership, LaVoy scowled at Cowsen and put his head back on the mound of soft dirt he'd gathered as a pillow. Cowsen snatched the disheveled packet of documents from Fricker and dropped it on LaVoy's head. Though the man stirred, his eyes remained shut.

"Hey, Sleeping Beauty, where are we?" Cowsen yelled. "How many men do you have? How many enemy are waitin' for you? When are you going to feed your troops, Ranger? You don't know, do ya, Ranger? He turned to the gathered mass of students and hissed, "The rest of you better keep awake, better know where you are at all times. You never know when LaVoy here's gonna die."

Though LaVoy roused a bit, he didn't look up, and Cowsen threatened, "I'm going to whup you upside the head if y'all don't get alert in a hurry." LaVoy's eyes actually opened, but he just lay there, dazed, until Fricker crouched and shook him. LaVoy looked up long enough to accept the envelope of orders, but as soon as Cowsen departed to roust the other dreamers, he nodded off again. So, Branch dropped to a knee, grabbed LaVoy by the fatigue shirt, and hooted, "Hey, boy, get your ass in gear. Sooner you issue orders, sooner I eat."

LaVoy sat up and created a plan, though his men were dismayed to learn that, while a meal was in the offing, as outlined on the sheet of paper Cowsen had surreptitiously slipped deep into the pile, the marching orders revealed they had to walk eight miles through the forest to get it.

After several hours, Fricker appeared at the head of the column on LaVoy's heels to grouse that he was dying of hunger and wanted a bowl of oatmeal. LaVoy, without turning toward him, spat, "Shove your stomach up your ass, Ranger."

As shafts of sunlight faded from the woods, Class 68-B was subjected to the second dictum of commando longevity—remain five meters behind the troop to your front. Any closer cost a whack across the face with a tree limb, one that was silent coming, but slapped so resonantly, it extracted rounds of snorting laughter from the cadres a quarter-mile away. If the victim had collided with a pricker bush, the most abundant vegetation in Georgia, and said flora struck the cadet's eye, there was the usual smack, the reflex laughter, and then an added period of banging, crashing, cursing, and finally silence as the blinded troop acquired the aptitude to see with his ears. For an hour, the wounded soldier was compelled to listen for the footsteps of the Ranger to his front. He learned to stop when the next man stopped and start again when the guide-Ranger's feet ground forward. Most importantly, the Ranger with the bleeding eyes had to prepare himself to swim out of the freezing ponds and waterholes into which he invariably stumbled, as if drawn magnetically.

The odd cadet who defied the unwritten rule that proscribed audible grumbles about facial hemorrhage and blindness was descended upon by swarms of cadres, packs of wild dogs, who goaded and gnawed until the man shrieked at the top of his lungs, "I quit this shit!" That cue sent the cadres melting back into the forest, all but one officer, who remained behind to wait for an ambulance, or if far enough out in purgatory, a helicopter, to remove the failure.

Fricker growled, "Listen to those assholes. They're laughing. They don't give a shit about us. All they care about is their damn coffee and sandwiches comin' out on the meat wagon." For the next hours, while awaiting the rescue transport, the Rangers spoke in hushed tones of the wretch, who had been ordered to stand apart from the group, facing a tree.

Solaced only by dreams of someday boarding an airplane for Viet Nam and being allowed to sleep for the twenty-six-hour journey, J.W. followed Branch's crunching footsteps through the pitch darkness. Branch trod on, groaning a refrain for hours, an entreaty to a supernatural being with whom J.W. was not familiar. "Sylvester," he moaned, "where are you? I need you, Sylvester. Say you haven't abandoned me, Sylvester."

During a stop in the march, when Platoon Leader LaVoy realized he had piloted his men up the wrong mountain, J.W. crawled to Branch and whispered respectfully, "Is Sylvester a god in your church?"

Branch ignored him for a while then snorted as the column began to track back down the hill. "Yeah, he's a god. Man, you are a damn fool."

J.W. shrugged his shoulders, turned to Fricker, and sighed, "Now what did I say?" But Fricker would not answer, for the air was so frigid it burned his throat. He spoke, finally, in hushed tones to J.W., his lips as motionless as a ventriloquist. "I can't move fast enough to stay warm, just enough to sweat. And then that shit freezes right to my back."

Near the Upatoi River, while walking a step to the side and behind Platoon Leader LaVoy, all the time grunting demands for food, Fricker hit a soft spot in the loamy soil. He collapsed five feet into a frozen stream, causing a crust of ice to shoot rearward, a shard of which coasted through a hole in the crotch of J.W.'s fatigue pants. Convinced he had been shot in the genitals, J.W. grabbed himself and screamed, "Sergeant Cowsen, help!"

Cowsen arrived shaking his head, "What the heck is it now, Ranger?"

But J.W. had since discovered the fragments in his pants were icicles, not shrapnel from a mortar round, and as he drew a breath to launch his explanation, Cowsen about-faced and moved silently through the brush to fish Fricker from the creek. A moment later, though, Cowsen abandoned Fricker in order to silence the platoon's point men, a pair of Rangers who were crashing through the deadfall back to the main body, howling at the top of their lungs they had spotted the foe just over the next ridge.

The two reported, with great animation, that they had reconnoitered a battalion of lounging aggressors, many of whom were asleep. "Didn't see a single enemy on security."

LaVoy rubbed his hands excitedly as he called for his men to quiet down and prepare for the attack. He directed Second Platoon to drop their packs in an open field to allow them to move faster and with greater stealth, then gathered his commandos for a final briefing and an attack rehearsal. This they did in slow motion, as if walking through a football play. Then he had his men feign firing their rifles, mouth whispered pop-pop sounds, and pretend to overtake and bayonet the imagined enemy. After running through the dummy drill several times, until all the men were grunting under their breaths like machine guns, he moved to the front of the platoon, pumped his rifle in the air, and took off downhill toward the objective.

At first, they busted brush slowly, carefully, protecting their arms from the thickets of pricker bushes, but LaVoy, quaking in anticipation, prodded his men to move faster, tantalizing them with the coming thrill of battle. At the crest of the next hillock, his scouts pointed out the clearing in which they had spied the aggressors.

LaVoy's eyes shrank into focused slits. Fist repeatedly piercing the air, he mouthed, "Follow me!"

In the dim moonlight, he dove forward down the hill, gathering speed, occasionally tripping on himself, all the time squealing obscenities at his men, instilling in them such fervor, even Morelotto and Fricker broke into a jog. When J.W. spotted Cowsen behind a tree, he nearly ruptured himself leaping into a sprint. Branch and Bearchild, however, had appointed themselves rear guard and walked very slowly, facing backwards, toward battle.

When the lead elements of Second Platoon spotted the clearing and the lounging enemy, they began firing blanks from their M-14s. With the blare of thirty rifles and machine guns pounding the forest, even the stragglers who'd lost their way were suddenly aroused. The entire platoon, sans Branch and Bearchild, blasted through the undergrowth zealously, soon exhausting their ammo, reduced again to making pop-pop sounds. But at the same time, they rolled with laughter, truly happy, the last time many of them would ever be.

They were greeted at the clearing by the shadows of enemy troops rushing about in disarray. The hostiles had been surprised beyond LaVoy's wildest dreams, and he ran through the enemy camp screaming at his underlings to fan out and subdue those cowards attempting to escape. While many of the conquered plopped down on the earth, arms lifted in surrender like so many exterminated beetles, a new problem emerged, for LaVoy had failed to include in his attack briefing appropriate stewardship of prisoners.

Fricker looked to J.W. for guidance, "We supposed to hit them in the head with our guns?" J.W. did not answer, for he was enjoying the battle zone of running soldiers and sitting soldiers, enlisted aggressors and officer Rangers, two populations with very different pasts and far different futures. Though dressed identically, he believed it was easy to discern his platoon from the opposition. As one of the aggressors ran tauntingly in front of him, J.W. whacked him so hard in the mid-section with a football forearm, the man puked. Second Platoon had fostered an unprecedented level of bedlam on the banks of the Upatoi, and J.W. stopped to drink in the pandemonium and enjoy the grunt of retching superimposed over the belly laughter.

Just about the time J.W. had laughed himself into lightheadedness, a fuming voice screamed, "Cease fire, assholes!"

The tenor of that utterance had the mark of a Ranger instructor, and it brought a dramatic silence to the war zone. A parting cloud exposed the moon, making it easier for the two sides to see the scowls they were flinging at each other. In the murky, blue light it became obvious to the combatants how similar they really were—identical faces, identical shaven heads, identical red-mud-stained uniforms, identical oil-stained hats. The aggressors were armed with M-14s attached to their pistol belts with bootlaces, just like the Rangers. It was hard for J.W. to believe the army dressed and treated the teachers' helpers as badly as they did Rangers.

"No," he shook his head and yelled to Flicker, "Something's wrong here."

"Assholes," the staff sergeant thundered. "Those aren't aggressors! You fuckin' idiots just attacked your own Third Platoon! Do not move."

Cowsen walked quietly to the center of the clearing then ordered calmly, "Fall in." He paced, his hand rubbing his chin. "Rangers, your field intelligence was lacking, not so much in quantity as in quality. There is a lesson to be learned here, Rangers." He paced a bit farther then looked back at the young men who waited in great anticipation for the pearl of wisdom, the distillation of a twenty-five-year career as a combat infantryman in three different wars, the single concept that would one day save them and their men. "What the heck you looking at, Rangers?" he blared. "You'll do it again, and then again, and again until you get it right."

The cadres dove into the puddle of troops, ordering them into squads and then platoons, but as the Rangers drifted about in the winter night searching for the faces they had known for but two days, clouds again enshrouded the moon, and the staff sergeant screamed, "I don't want no cluster fuck, Rangers. You're going to be very sorry if you don't form up quick like. Do you hear me, Rangers?"

The class came to attention where they stood, brand new squads and platoons taking form in the pre-dawn minutes of the Fort Benning outback. Second Platoon, now with several hesitant, but silent, Third Platooners interspersed, was marched up the hill toward the clearing in which their heavy gear had been dropped. Though LaVoy was absolutely sure he had led them back to the right spot, the equipment, every stick of it, had disappeared.

LaVoy walked to the center of the clearing and sat on the ground, head hanging. Cowsen stood above him and spoke softly, "Ranger, you got a lot on your plate. You need to find your equipment, redeploy into attack posture, overrun the enemy, try the real one this time, and then sweep past the objective and mop up straggler aggressors. And maybe," he smiled, drawing a tuna sandwich from his rucksack, "you should consider chowing down your men." Cowsen started to walk away but called over his shoulder, "And you'll do it all before 'oh-five- hundred'." And that's less than three minutes from now, Ranger."

LaVoy was frozen to the earth by the weight of his world, but he eventually jumped up and called out, "Rangers, I have chosen to attack. We must accept that our gear is gone forever. For the hours we'd spend tripping over deadfall searching for it, we could be done with our mission and on our way to breakfast."

Fricker raised his hand. "My mess kit's in my rucksack. What am I supposed to eat out of, my hat?"

LaVoy ignored Fricker and also the entreaties of a growing legion of grumbling Rangers. He set out down the hill toward the new objective. Cowsen caught up with him, banged him on the head with a small log, and asked, "Are you going to brief your men before the attack, Ranger?"

LaVoy called the troops together for a fresh pre-battle conference. He shoved them into attack formation then insisted upon a rehearsal or two, holding them there until the "pop-pops" were broadcast with sufficient enthusiasm. Then he started with the-arm pumping again and took off down the hill. Fricker and J.W. wandered behind, their minds kilometers away as they chattered about gritty, stale GI coffee. J.W.'s eyes glazed over. "You know those little packets they put in Cs? Man, I can just taste it; bitter, yeah, but I'm gonna fix that right up. Watch me now." He pulled from his fatigues a single packet of water-logged sugar he'd liberated from the cadres' table back at the mess hall.

Fricker's jaw dropped. "You'll divvy it up, right, man?"

"We'll see."

Fricker gave him a little tap on the shoulder. "Can't you just picture it? Lounging against a tree, sharing, warming our hands around a steaming canteen cup of java. Man, that's what friendship's about." Occupied by the illusion, the two reached the bottom of hill but took a wrong turn and stumbled into Third Platoon's staging area. They were captured and gagged.

Minutes later, the two platoons met again at the bottom of the hill. Prisoners were exchanged, new orders issued, and the business of commando school revived. LaVoy, despite arming himself with a fresh plan, misread his map, and the end of the hour found them even further from the river than at the start of the original attack.

The new platoon leader, Gillette, made a quick map assessment, laughed derisively at LaVoy's ineptitude, then drove the platoon back up the hill into the clearing where LaVoy had stopped to brief them the second time.

"Seats, gentlemen," Gillette bid with relish, and his men collapsed on the frost-laced forest floor. Bearchild's face became peculiarly sedate, and when Branch looked at him inquisitively, Bearchild whispered, "I'm pretending it's a pow wow."

"Gentlemen, my name is Ranger Gillette. I am your platoon leader. I am going to brief you on the plan I have devised to capture our objective."

Branch's eyes rolled up in his head. "Who gives a rip about a plan? I need something to eat. I could be sleeping, and this jerk off's worried about an objective."

Morelotto had the courtesy to raise his hand. "Hey, Ranger, are we gonna chow down or not? I'm hungry as hell."

"Fellow Rangers," Gillette began, "We have a mission to accomplish. The status of your appetite is secondary. With the help of God, half of the Second Platoon will ford the river, flank the enemy on the far bank, signal the troops on this side, and together, we will strike in a double-pronged attack. Those remaining on the far bank will clean up the enemy and capture prisoners as they run from us. Gentlemen, if this is done properly, they will be taken by complete surprise."

LaVoy raised his hand. "You outta your mind? That river's like ice. Let the enemy swim in it. Let them freeze their peckers off. That'll slow 'em down."

"The enemy will never expect Rangers to work that hard," Gillette proffered condescendingly. "Are any of you aware of Vo Nguyen Giap?"

Without raising his hand this time, Morelotto offered, "Man, I don't give a shit about no Jap. World War Two's over, in case you didn't hear. I need some food and some sleep."

Bearchild raised his hand. "North Vietnamese general. Defeated the French at Dien Bien Phu in 1954."

"Precisely," Gillette smiled approvingly at Bearchild. "That's what I'm trying to tell you. He motivated his men to do the impossible in the mountains of central Viet Nam. The French never dreamed humans had the will to fight so undyingly for independence.

"Independence, gentlemen, had not meant much to the French during World War Two." His hand waved in a grand professorial arch over the stump upon which he had piled the worn mission orders. His eyes drifted toward the heavens, transported back in time, reliving history, relishing a lecturer's bliss. "Yes, back in France, a few of them became partisans, indeed, but a boundless number of the Frogs pandered to their Nazi masters. On the other hand, less than ten years later, in the French colony of Viet Nam, the entire North Vietnamese army was willing to die to kick out the French occupiers. They hauled tons of artillery up near-vertical gradients no matter how badly it hurt, no matter how many men had to die doing so. And because of the Vietnamese bravery, the invaders were sent packing all the way back to France, tails between their legs. Are we the French, Rangers, or the Vietnamese?"

Morelotto, who had fallen asleep, tipped over onto Fricker, who fell over onto the legs of the staff sergeant, who lost his balance, stumbled to the ground, and banged into Cowsen.

"Ranger Gillette, get this goddamn patrol on the road. Now!" the staff sergeant howled, brushing off his mulch-stained fatigues.

When the men began to move, Cowsen ambled up to Gillette and asked, "Ranger, where'd you learn all that?"

Gillette offered brusquely, "Harvard, First Sergeant."

"They lettin' any old black sergeants in there, Ranger?"

On the march to the Upatoi River, a circle of malcontents met during a three-minute hiatus. Morelotto spoke first. "Man, I say we delay. I mean waste as much time as possible. If we don't cross until dawn, sun'll be up. Warm the river. Fuck Gillette," he sneered, "I hate them Harvard assholes."

Branch smirked, "Sylvester ain't gonna be warming the river none, fool. He'll warm the air, not the water. And man, is it worth a tenth of a BTU to screw one of your own? You white people really are somethin'."

Morelotto turned to Bearchild for support, but Bearchild drifted off toward Gillette. When Morelotto petitioned J.W., Weathersby mumbled, "I don't know, man," then walked off toward Branch.

Soon after moving out, Gillette dropped back and gathered his men. "I've just been given orders to be across the river by sunrise. They're going to have Cs for us on the other side." Morelotto, though, had drifted off in a stupor and was missing. The rest of the platoon petitioned the staff sergeant to let them run to the river and leave Morelotto to his own devices, but he refused to let them go on until Morelotto was found. Several more Rangers became lost during the rescue mission, and by the time all the troops were accounted for, the eastern sky had lightened ever so slightly with an orange hue.

In the end, Morelotto's ploy to delay the progress of the mission was for naught. The first shafts of winter sunlight were little warmer than those of the setting moon. His ceaseless muttering, and Branch's never-ending supplications to Sylvester, served only to burn calories they didn't have.

Clean-shaven and in pressed fatigues, Cowsen met the platoon by the river, as chipper as if he had consumed a country breakfast of pancakes, bacon, scrambled eggs, a side of hash browns, three slices of buttered, white toast, and several cups of freshly brewed, deep roast coffee, then driven back for a new day at the office. He scrutinized the lot of young commissioned officers milling about the banks of the Upatoi until happening upon a target of opportunity—J.W. Weathersby.

"Get your butt over here," Cowsen chuckled. "You're the one who can swim forever, right? Take off your fatigues, Ranger."

"Sergeant Cowsen, right here, sir?"

"Don't call me sir! Yes, right here."

"I didn't think you cared," J.W. tittered, but Cowsen's eyes hardened, and Weathersby unbuttoned his ragged fatigue jacket.

Cowsen announced loudly, "You, Ranger, are christened the company swimmer. You are designated to forge water obstacles from this dawn forward 'till Last Judgment. Like I said, take off all your clothes." Cowsen's eyes opened, black, wide, searching until he nodded in satisfaction and walked behind the gnarled trunk of an ancient oak. The Ranger who had been punished and forced to carry the thirty-pound climbing rope for sleeping during class was on his back, snoring, oblivious to the shards of ice Cowsen was dropping into his open mouth.

The master sergeant hovered for the first few of the fifty-one push-ups then yawned and pulled from the pocket of his crisply starched uniform a tiny reel of fishing line. He handed one end of the thin nylon thread to J.W. "Tie this to your ankle, Ranger, then hurry up and swim across the river. If you make it, Ranger Branch will tie this here three-quarter inch hemp rope to the fish line, and you will pull the whole doggone thing across to the other side so the rest of your troops can get across safely."

Cowsen's eyes dropped to J.W.'s drawers and socks. "Ranger, I thought I told you to strip. Give your clothes to Branch here. He's gonna wrap 'em in *your* poncho and ensure that your gear gets across nice and dry. Aren't you, Ranger Branch?"

Off to the side, the staff sergeant squatted disinterestedly at the Upatoi's edge, filling his canteen cup with bits of crusted ice and water to make coffee, J.W. fantasized, but as J.W. hesitated to spring from the bank into the river, planning to suffer the cold inch by inch at a tiptoe, the staff sergeant tossed the ice water across his back. J.W. shot forward, instantly paralyzed by the cold. Try as he might, he could not make his arms tread against the freezing, brown current.

Cowsen watched J.W. drift downstream. "Better start a'swimmin', Ranger. Y'all be floatin' into the Walter F. George

Reservoir if'n y'all don't get them arms a churnin'. Pretend yo arms is yo mouth."

For a few seconds, J.W. swam like his departed Ranger buddy had climbed the cargo net, a man possessed, but his arms and legs quickly numbed in the frigid river, and he began to float with the current downstream again. He pictured Smith, moribund on the ledge, and realized it was now his turn. He wondered if he, too, would be forgotten as quickly as his first Ranger buddy. But Cowsen screamed again, and J.W. kicked with a mighty effort that propelled him toward land. At the far bank, he scratched madly at the freezing mud to lift himself from the water, then dropped onto a rotting tree trunk to rest, breathing as heavily as if he had carried that log across the Upatoi on his back. Cowsen harped from the other side demanding J.W. speed up, so Weathersby reached down to his ankle to untie the nylon fish line. It was gone, pulled free, he guessed, by his mad flailing.

Cowsen hollered, "Turn around, Ranger. Get the line back."

J.W. crawled numbly into the water, but in an instant, his arms turned to lead and, blinded by the blistering, cold sun clearly rising in the east, he began a slow drift downriver, no longer caring.

Cowsen threw several rocks at J.W., who gathered the energy to dodge them, though the last one grazed his head. Perhaps it was the pain or the cracking sound, he was not sure, but he found himself suddenly awake, struggling upstream toward the ripple on the surface of the water created, he was sure, by the nylon line. As he relaxed momentarily, marshaling the strength to reach out and grab the line, he noticed a small, light green smudge in the roiling waters near his hand. He looked harder and was sure he saw a pair of dripping fangs nestled in a field of white. It took a moment to partially decode the message, for few neurons remained sufficiently warm to fire, and the best he could reckon was that the ripple had to be a massive, poisonous anaconda. J.W. shrieked, "SNAKE!" as if yelling, "PASS!" as a linebacker at Sterling College. On his football team, several of his mates surely would have sprinted to back him up, though that early morning, no rescue party was dispatched. In fact, no one on the bank bothered to look.

Cowsen called over the rush of the nearby rapids, "Drive on, Ranger. Swim upstream and get the doggone line back. Don't pay that garter snake no heed. He'll be movin' on shortly."

And indeed, the current of the Upatoi carried the snake past J.W.'s face, the serpent's white jaws opening and closing rhythmically in tune with the violent slapping of Weathersby's hands.

Soon after struggling up the far bank, he was ordered to start pulling the fish line and drag the waterlogged hemp across the Upatoi. Sweat mixed with the foul river water made his hands slick, and with the staff sergeant's reverse tug on the fish line, the process was delayed for another hour, long enough for the Rangers to miss their rendezvous with the mess truck.

When J.W. finally grabbed the actual rope, he discovered it was long enough only to reach the edge of a single, rotted tree, one barely inches from the waterline. He stood on the far shore in the buff, shouting for Cowsen to play out a bit more rope. With no reply, and running out of breath, he settled on tying the hemp to the decayed trunk poking from the bank. The knot he wove was enormous and slovenly. He blamed it on the numbness in his fingers, dismissing that he had slept through the lesson on ropemanship the day before.

He was surprised when the rope held Branch, the first to monkey climb across the Upatoi. The small, muscular man moved with the same power he had over the gulch at the obstacle course and on the wire at Victory Lake. Bone-dry, he lobbed J.W. a sloppy, basketball-sized parcel of dripping, river-scum-soaked fatigues, socks, and boots.

Fricker, Bearchild, Gillette, and even LaVoy crossed without incident as J.W. husbanded the rope, tightening it, repositioning it, taking care and some pride in having been so an integral part of the crossing. When the last man, Morelotto, plunged onto the rope, his two hundred and thirty pounds strained the wet hemp until the anchor stump cleaved from the bank, dropping the primitive bridge into the drink. Nonetheless, Morelotto held fast to the line, twisting in the muddy current of the Upatoi, a harpooned, olive drab, inland whale. While aiming a gurgling string of Mediterranean epithets at J.W., he swallowed gallons of river water, and it was some time before, at Cowsen's urging, Morelotto turned his attention away from retribution to concentrate on a struggle for land. Morelotto pulled himself back to the near shore, landing directly under Cowsen, who was resting against a tree, picking his teeth with a weed.

"Do it again, Ranger, until you get it right. And this time, you take the rope across."

Despite Morelotto's inflammatory mien, and J.W.'s penchant for self-blame, Fricker reassured Weathersby that not one of them had taken courses in college about frozen tropical rivers or garter snakes. "It's not your fault the tree trunk was rotten and the rope didn't stretch to one of those good trees. Don't worry about it, man."

"That's right," J.W. answered with bravado. "He can kiss my ass. I didn't ask to be in Ranger School in the first place."

But when Morelotto eventually reached the far bank, he whispered threateningly, "Your ass is mine, chump."

J. W. set off on a fifteen-minute apology, ending with a promise to never let it happen again.

The two squads Gillette had charged to form the pincer movement across the river shuffled into attack position and waited for the balance of Second Platoon's men on the near bank to initiate the battle. The two halves of Second Platoon struck within seconds of each other, closing the vise. The Green Beret aggressors dropped to the ground in perfunctory fashion, though rather than feigning death, they leaned against trees, smoking fresh cigarettes and drinking steaming coffee from thermoses.

Gillette strutted about the battle zone, jabbing his fist forward, grinding his jaw, "Damn, that was perfect! Man, that was good."

When he jigged past Cowsen, the first sergeant kissed him across the back of the head with the magic stick. "Ranger, you're dead, choked to death on a silver spoon."

He then thwacked the wand on the bald spot of an older Ranger who had taken a break to doze under a bush. Cowsen growled to the new platoon leader, "Choose a rallying point. Designate a spot on the map where those who survive the attack rendezvous, then regroup, and you need to care for the enemy wounded. And when you get all that easy stuff done, plan the next mission." Cowsen became even more solemn and warned, "The rallying point you choose, Ranger, it better be one the enemy can't find in a thousand years. I'm giving you fair warning."

The new platoon leader studied the map and the instructions for half-an-hour, wrote formal commands, then announced he had found a perfectly secure site, one even battle-tested Green Berets recently back from Viet Nam could not find. He carefully laid plans for his men to eat and rest at the rallying point, a promise that brought cheers and a grand whoop of delight when he declared, "Gentlemen, our objective is only one kilometer from here. That's less than a mile as the crow flies!"

It took an hour to find the clearing, and when they arrived, it was piled six-feet-high with their lost battle gear from two leaders ago. As Second Platoon dug through the heap, Fricker looked about and babbled, "Hey, that Cowsen guy's not around."

Gillette surveyed the clearing and pronounced, "A serendipitous absence." The new student leader, however, had not yet devised a plan to continue the attack after he brought them to the clearing, and in the absence of orders and, peculiarly, of cadres, he stood to the front of his platoon, confused and silent.

Morelotto called out from the ranks, "Hey, man, I say we go administrative."

The entire platoon grunted in concord and, without waiting for ratification from their leader, dropped in mutiny to the ground. By the velvety feel of the twigs, J.W. surmised they were near water. When the crunching of soldiers collapsing on the deadfall ceased, and all the men were comfortably curled in the fetal position, J.W. heard the burble of a nearby stream and the churn of the Upatoi's rapids in the distance, sounds that dampened the pain of the penetrating cold. Soon, he sensed a peculiar glow in not having to move. Though J.W. noticed his feet lying in a puddle of icy slime, he was so comfortable, he chose not to lift his legs, even when the water wicked slowly into his leak-proof boots, up his socks, and finally into the hem of what was left of his fatigues. While he could see the osmosis with his eyes, he could only imagine the pain of the icy water on his skin, each sheet a degree colder than the last, for his feet had, hours before, become anesthetized.

According to the map, they had trekked a dozen miles since sunset, but J.W. was oblivious into which dimension the hours had vanished. Forgetting that he was on the ground, his only memory was of the incessant crack of twigs and the thousand boughs of cedar and pine that had cuffed his face along the way. When he heard the rustling of undergrowth becoming louder, an uncomfortable twinge shot through his chest. He panicked that it might be time to resume the death march, but it was only Branch skimming about the clearing, gathering dry stalks and bits of grass.

Branch dumped his pickings alongside chunks of hefty wood then dropped to his knees, struck several matches before one lit, and started a tiny fire fueled by the blades of grass. He tended the flickering yellow flame like a starving man worshipping a granule of food, reverently adding shreds of grass, one here, one there. He blew gently on the candle-sized fire, and as a meager flame built, he tenderly took the thinnest twigs, little more than grass themselves, and added them discretely to the tiny, smoking cone. The tinder glowed cool red at first then matured into a warm orange. Branch looked up confidently, a smile of triumph and excitement gracing his pleasant features. He seemed not to notice the flame degenerating a shade or two deeper, back into the red hues.

Bearchild observed the darkening embers. He crawled behind Branch, waited as long as he could contain himself, and as the fire neared demise, pushed Branch out of the way. He puffed ever so gently on the remaining sparks, slowly culling a glint of orange back to the surface. He slowly added slightly larger twigs until a tiny yellow flame struggled out of the embers. He rubbed a few strands of wet straw on a dry patch of his fatigues then added them, one by one, until the flame brightened. Branch tried to drop a few larger pieces on the barely glowing enterprise, but Bearchild sidled around on his knees, blocking his Ranger buddy. Bearchild blew harder, carefully situating a few more tufts of grass and moss on the coals. The yellow brightened.

Branch's eyes glazed over, a man possessed with an insatiable desire to be warm, to touch the flames. He pushed his way around Bearchild and placed larger twigs on the fire, but Bearchild picked off the added fuel with callused fingers, replacing it with shafts of grass only a micron larger than the last. As the flame built, the smoke dissipated, and Branch's face relaxed as milli-BTUs radiated from the birthday-candle flame. Bearchild added more grass, each blade a fraction larger than the last. He picked through his stock of kindling and chose, with some deliberation, a twig so thin a toothpick would have gloated in its own girth. His blaze now rivaled the flame of a fuel-starved Zippo lighter.

With the contemplation of a symphonic conductor, Bearchild chose for his next movement only subtly larger and barely more impressive kindling, twigs as hefty as Q-tips. Still, he had not uttered a word. No one had. Instead of speaking, the soldiers crawled together, gathering in silent prayer, waiting expectantly for a Rose Bud Sioux Indian to deliver them.

But the wet tinder collapsed into dying red embers, and some of the troops sighed and closed their eyes to sleep. Bearchild, though, did not. *His* eyes focused intensely, his jaw set even more tenaciously. He began again with microscopic chaff then puffed until a flame appeared. He deposited the driest of twigs tenderly on the coals, gently blowing life back into them. As they ignited, Bearchild cautiously added sticks and then thin boughs. Just four minutes later, the outsides of several small logs were glowing. Two minutes after that, he posited a rotted tree trunk on the pyre. The air seared with a heat so fierce, Branch backed off several feet. Bearchild, however, held his ground, the pilot of a great engine.

With the flames sustaining themselves, he sprang back into the forest, howling incantations and ripping rotten trees wildly from the tundra. When he returned to the clearing, he slowed majestically and placed each new log onto the blaze with great decorum, as if he were delivering a newborn to its mother's arms. When those timbers caught, and the platoon encircled the inferno, Bearchild stopped. His face relaxed and it seemed he would join his comrades in celebration, but he dashed back into the woods, harvesting more salvation. He returned over and over to the fold, flinging wet wood, dry wood, hunks of every natural fuel in the Georgia outback, onto his masterpiece. He worked as if the sun had exploded, leaving only hours on the clock for all of mankind, believing it was his charge to warm the entire nation and save every freezing American from the curse of cold. When he finally stepped back, the forest glowed as if an entire apartment block had been set ablaze.

Within thirty seconds of Bearchild's triumph, however, the woodland rumbled and Cowsen emerged from the tree line, walking slowly, a coldblooded snake slithering toward a comfortable, hot rock.

"Rangers!" he called out, "no order was given to start a fire." The few cadets not inert from their pleasure crawled closer to the heat. "Every man will deploy his entrenching tool and bury that excuse of a birthday candle. Ranger Branch, did you light that thing? You Rangers are going to learn this ain't no picnic. This ain't college no more. This is the real thing, Rangers. Do it. Now!"

Second Platoon shoveled slowly, scraping peat and kindling off the forest floor to sprinkle onto the blaze. That just fed the monster, so Cowsen snatched Branch's shovel and excavated through the soft forest detritus, uncovering a layer of red clay. He chucked lumps of the dripping earth upon the blaze. In half-a-minute it was all over, and as the last ember expired, the platoon of Rangers began to shiver in the cold.

Cowsen stood with arms folded across his uniform, warm, free of river water. J.W. mimicked him and noticed the front of his own fatigue jacket had dried in the heat. He turned to Fricker, "Young trooper, I hereby swear to preserve the ecstasy of dry warmth, no matter the cost. And with that declaration, I begin Day Three, or is it Four, of Ranger School."

Cowsen drove them through the forest, double-timing the clearings, power walking the wind-whipped timberland. It was not long before J.W.'s body temperature drifted back up into the low one hundreds, and the front of his fatigues dripped with sweat. In fifteen minutes, his entire uniform was sopping wet, all except the hems of his pants and fatigue jacket, which were again frozen solid, caked with rime ice.

At noon, Cowsen drew the magic stick from his pack. He walked a curious path a fair distance from the Rangers, who backed away and broke toward the brambled forest to hide. There was earsplitting vulgarity as the troops were ribboned on the prickers, though Cowsen ambled tranquilly along a path free of vegetation, examining each of the mired cadets until he found his mark. A bald head was snapped with the flimsy baton and the orders were handed over, but not before several sheets blew away in the wind. The change of command terminated with Cowsen disappearing into the woods.

The latest student chief called a formation. "Relax, gentlemen, Sergeant Cowsen assures me we have breakfast to look forward to.

A man can endure just about anything with a hot meal simmering at the other end. Hang in there a little longer, and this whole thing'll be over. It says so in the orders. Our next objective is the mess hall at Harmony Church"

He thrust the manila envelope into the air to shouts of "Drive on," "Hot damn," and "Victory!" With a howl of hope, Second Platoon dashed toward base camp for the beginning of what was rumored to be the end of the breaking down process and the commencement of the rebuilding. The skinny was that meals would be nutritious and plentiful; Rangers were from that point forward to be treated like the elite soldiers they had been selected to become. J.W. sprinted near the head of the pack—finally, something to live for.

At main camp, LaVoy was sent to stand at attention in front of the blackened, wooden door of the command shack. J.W. spied him, eyes cast down, as the rest of the company was driven at the double-time toward the pull-up bar. And then more troops were plucked from the formation during the run, the staff sergeant goading them toward the shack. The ten outwardly despondent washouts were herded into the mess hall for a brunch of leftover, greasy bacon; watery, lumpy, mashed potatoes; and warm milk.

The remaining Rangers fell into company formation, one riddled with not a few gaps. They instinctively closed ranks to make up for the missing mass of the defeated, a schema of military orthodontia in which every cleft was swiftly filled.

They double-timed to a set of bleachers nestled in a narrow valley between Hamilton and Reeve Hills. The freezing gusts bit through the company of Rangers who sat there for the day, exposed to lectures, not one word of which J.W. absorbed. When the sun set, and the winds died, those who had survived the first tactical mission moved on the double toward the mess hall, the staff sergeant calling cadence, grinning through clenched, smoke-stained teeth. When they sailed past the pull-up bar, most smiled at the promise of dining without calisthenics, but they also sailed past the front doors of the mess hall. Sergeant Hartack had them loop around the mess hall, twice, then sprint back into the field.

Straw men had been erected for the class's second visit to the bayonet course, faux victims at which J.W. lunged viciously, driving the foot-long blade poking from the business end of his M-14 into their cellulose guts. With each stab, he bridled more furiously in contempt for the platoon leaders who had failed their missions and cost him food and sleep. That the most bungling of them were gone was no consolation; it only deepened the rage. His fellows joined him in tearing violently at the mannequins, spitting on them, cursing them, so incensed, all they left behind was shredded fodder and the reverberations of violent anger.

When they force-marched back to Harmony Church, the staff sergeant croaked, "Rangers, the ante's been upped—twenty-five and one to be granted entrance to the hallowed hall of mess. We don't want to see none a youse gettin' fat."

A new sergeant stood guard over the mess line. "Formation in five minutes," he badgered as J.W. picked from the smorgasbord of cold eggs, cold bacon, watery mashed potatoes, stale Wonder Bread, warm melon balls, and tepid milk. J.W. piled his dented metal tray with the offerings, scarfing the bacon before putting his meal down on a table far from the staff sergeant. As he raised his head to allow a lump of potato down his throat, he eyed Gillette, surrounded by a fire team of Rangers, bowing heads and clasping fingers in silent prayer. J.W. managed to gobble several forkfuls of mucousy eggs before the worshipers looked up, their faces as serene as a circle of Buddhist monks undergoing self-immolation. J.W.'s fingers, on the other hand, moved even faster, hurling strips of congealed bacon into his mouth.

As the oily pork slid past J.W.'s pharynx, his esophagus spasmed, a semiconductor, incapable of swallowing, but not of urping up thick jets of a bitter, acidy, chunky emulsion reminiscent of that which had filled his throat during the fall into Victory Lake. Branch raised his eyes to search for the source of the choking cough, but when he discovered it was just J.W. Weathersby, he went on packing his face.

Slowly, the throbbing in J.W.'s gut evened, the pain replaced by the sensation of a fifty-pound slab of anchored granite. He could not consider another mouthful, and he left his seat early. As he bent forward to dump the balance of his meal into the waste barrel, he burped up more acid. That resurrected the hacking.

As the last dribble of mashed potatoes slithered off his tray into the garbage tub, the mess sergeant, alerted by the din of J.W.'s coughing, bristled over to demand an explanation. "You know how many hours of toil it took to prepare what you just discarded, Ranger? Ranger, what the hell's wrong witch you? Throwing away food! Throwing away food? Why, in few days Ranger, you'll trade a piece of ass for half a peanut butter and jelly sandwich. You mark my words." J.W. tried to respond, but cramps doubled him over. In seconds, Cowsen materialized. He gaped into the barrel, tightened his lips, conferred with the mess sergeant, and sighed. "Ranger, there are countless hungry folk in this world. You just don't care, do you?" His voice rose. "Ranger, I told you a long time ago, this ain't the Bronx. In Ranger School, you ain't no spoiled rich kid no more. Get your food out of that can, put it back on your tray, and eat it."

J.W. glared into Cowsen's eyes. "No, First Sergeant, I won't. That's it. I'm not eating out of garbage cans. This time you pushed too far."

"You'll eat it, or your butt'll be sittin' next to LaVoy and them on the bus. Take your pick." J.W. stood his ground. He could not help but hear the utter silence in the mess hall, and he realized he was again the epicenter of tumult. His eyes bored deeper into Cowsen's.

With his front teeth, he scraped from his tongue the yellow-green scum that had been incubating for days. He blended it with bits of half-chewed scrambled eggs, bacon, and the stomach effluent that continued to trickle up his gullet. He balled the wad and aimed for Cowsen's lips, smiling, imagining the blend dripping from his tormentor's chin. As the slime gelled just behind his teeth, he also imagined the thud of Cowsen's fist against his mouth.

The nuclear standoff went on for a good three seconds before Weathersby's shoulders drooped. He leaned forward, dipped into the can, pawed at the swill, squeezed a handful past his lips, and mixed it with the ball of debris in his mouth. He allowed himself to start back into the position of attention, though, despite his best purpose, he gagged in Cowsen's face. Both the first and mess sergeants, as if they had suffered this scenario once or twice before, lurched backwards, but J.W. retched nonetheless, and the two non-coms were shocked at the speed with which the salvo flew from his mouth.

Perhaps what saved his career, and his life, was that, just before the orb flew, he twisted his neck furiously to avoid Cowsen's released fist. And so the clump arched like a Pershing missile as it rocketed back into the garbage can. J.W., drooling slop, unbent to a position of attention. With bits of food flying from his mouth, he hollered, "Yes, First Sergeant. Won't happen again, First Sergeant."

Before Cowsen could seize J.W.'s fatigue blouse, Weathersby dashed from the mess hall then broke into a crafty jog toward the barracks. He slowed to look furtively over his shoulder to witness the staff sergeant discussing with Fricker the value of pull-ups in the rehabilitation of the human shoulder.

J.W. made it as far as the command shack before Hartack called out, "Back here, Ranger. Execute twenty-and-one on general principles, and then eleven more for a failed escape attempt." Half-way through, Hartack demanded, "What you smilin' at, Ranger? You like 'em or somethin'?"

"Actually, Sergeant, it stretches my stomach. Makes me feel better. Kinda like an after-dinner mint."

As the sun set, Cowsen gathered the company. "Rangers, this is a significant night. We will be spending it at the machine gun range. You will learn more about the M-60 machine gun," he advised serenely, "than you ever knew there was to know."

A hum of excitement spun through the ranks as the decomposing tip of the nation's spear reflected upon the prospect of being taught to use the .30 caliber machine gun. "Hey, man, I can't wait. Finally, some fun," J.W. hummed to no one in particular.

"I'm ready for a rest. An easy night." Fricker smiled.

Cowsen, though dozens of yards away, cracked, "You think so, Ranger? The only easy night in Ranger School was last night."

All J.W. would learn in the earlier reaches of that cold evening was how to carry the nearly twenty-five-pound weapon on the run. At the machine gun range, instead of setting up the guns and firing madly into the night, he and his select group of munitions bearers turned their cargo over to a gallery of angry, senior sergeants, more than one of whom sported long, deep, facial scars. Though these men had shaven only hours before, they had stubbles that would have taken J.W. three days to breed.

One of the machine gun NCOs grinned with nicotine-stained, gold and silver-capped teeth as he placed the weapons with loving care onto a line of tripod mounts. Only the very tips of the guns, however, protruded beyond horizontal steel pipes running barely an inch above and below the barrels.

Fricker whispered to J. W., "Shit, that's not gonna be any fun. I wanna do sweeping machine gun fire, like in *Merrill's Marauders*. You ever see that flick? It was about Rangers."

J.W. grumbled, "I don't give a shit about movies right now. It's those sergeants. Man, they look like pirates, like fuckin' gangsters."

"Yeah, and why do those bottom pipes have a V shape in 'em?"

"I donno. Must be the gun's name is Victor."

"All of 'em have the same name?"

"That's enough questions," a gruff voice declared from inside the first cage. "Welcome to the machine gun range, gentlemen. I am Sergeant Poliak. You will crawl the machine gun course flat on your belly. The rounds will pass eighteen inches, more or less, above the ground, directly over you. These here pipes here," he tapped the downward bent, V-shaped metal under the muzzles with a swagger stick, "are here for your safety. Do not be concerned. If you have lost the weight you should have, you will not be in danger."

J.W. measured the depth of various body structures with his hands, finding the most prominent area to be his butt, nigh sixteen inches front to back, he estimated. Then he spread two inches between his thumb and index finger, added it to the space he'd gauged, and raised his hand.

Poliak ignored J.W. but called out to Morelotto, "Your fat ass is mine, son of a big shot."

The other sergeants laughed hysterically, the cue for the first Rangers to be sent, one by one, crawling along serpentine, dusty paths. One of the first cadets laughed haughtily, "This ain't nothin' compared to the obstacle course. It's a fuckin' joke." And it was, until a cacophony of sharp explosions and the staccato of dazzling starbursts from the muzzles of four, fire-spewing barrels greeted the first of the scrambling soldiers. Four jumped up and ran back, refusing to go on.

J.W., Fricker, and dozens of other Rangers screamed, "Ceasefire, ceasefire!" but the blare of the guns drowned their pleas. With a break in the shooting, the whole class stood, tensed, fists clenched, while the resigning Rangers ran back chaotically toward the starting line.

Cowsen shook his head cheerlessly "What is wrong with you, Rangers?"

One man stood up to him. "There were bullets hitting the ground directly in front of my face. I saw them. They weren't eighteen inches off the ground, First Sergeant. I'm not going to be murdered here like that Negro enlisted man."

"Yeah," another man added.

Cowsen rolled his eyes. "That was the dry run. They were using blanks. The real bullets haven't started yet. Get back out there, and just for that, the rest of you Rangers, drop and gimme twenty-five-and-one for the Rangers of old. And you'll do them again and again until these few, fearless Rangers finish their business."

The four cadets dropped their heads and shoulders as they shuffled back to the course. Cowsen pointed at J.W. and motioned for him to join the team of cowards.

When J.W. came to freshly created hillocks of red mud and concrete, manmade topography that rendered it impossible to stay within eighteen inches of the ground, Poliak laughed, "Speed bumps. Slide on your back, Rangers, and keep your heads turned to the side—that might save your nose." When Weathersby did so, he noticed Cowsen throwing little stones in front of the Rangers' faces.

Halfway through the second dry run, the air began cracking with the supersonic report of genuine slugs. They were far louder than the blanks, and some of the troops jumped up precious inches, startled by the incessant hissing over their heads.

J.W. commanded himself to garner the nerves of steel required of a hero in training. "Drive on, Ranger," he growled to himself. "Keep crawling!" Though the message got to his fingers, which pulled at the dirt as they had in the underground tube, the rest of him did not budge. He lay shivering as the thick, blue smoke and acrid scent of red phosphorous tracer rounds condensed into a reeking soup.

But then the scent of cordite, spent gun powder, drifted over from the firing range. It sparked in J.W. an excitement, for it stirred the memory of childhood war games, of bunkers fashioned from cardboard refrigerator boxes, of sticks for guns, and firecrackers for artillery. The reek evoked one mock battle in particular, the time in his early teen years that he and his platoon of pals were pinned down. For a reason he never understood, he culled the valor that summer afternoon to arm himself with a pocketful of firecrackers and jump out of his cardboard barricade. In the face of a barrage of enemy sparklers, he fought his way along the hot tar of his homeland, Morris Avenue, to attack the invaders from the Grand Concourse. Crawling up to the enemy's cardboard box, he hurled a cherry bomb. When the battle was over, J.W. was awarded the Morris Avenue Medal of Honor.

Overwhelmed by that olfactory memory, J.W. at first navigated the course in a cautious slither but was soon sprinting on his belly, sweating, smiling, cursing easily as he passed the next Ranger, and then the next. As the stream of searing lead cracked closer overhead, J.W. fought with himself to keep from thrusting a hand into the air to touch it, to see if it was real. Instead, he flipped up a small rock, then a second and a third, until one was pulverized by a .30 caliber round. Dust blasted into his eyes and mouth, and he saw poorly, forced to feel his way along the circular path, which eventually deposited him directly in front of the guns.

Despite his partial blindness, he could make out, through the blue clouds of cordite, Sergeant Poliak's silhouette firing from behind the glaring, white muzzle blast. The barrel of Poliak's weapon glowed red, and J.W. stared, waiting for it to droop from the heat and send the trajectory of the bullets two precious inches lower. As for the sergeant, his eyes were wide, his lips pursed. He banged the barrel of his machine gun against the bottom pipe so hard, sparks skittered out. In his burning zeal to lower the barrel a few more sixteenths of an inch at the tip, he had visibly deepened the V in the safety bar.

Several more guns were suddenly manned, and so began a mad spasm of detonations that shook the ground and split the air. Sixty seconds later, someone called, "Ceasefire," and as abruptly as it had started, the bombardment ceased.

Poliak exited his cage and laughed, "Those of you who deferred on the first opportunity to experience the course are given one last opportunity for redemption."

The troops who agreed to reinstate themselves did the course twice with live fire, and once at the end with blanks. When the firing ceased for good, and the men enjoined to fall in, several spontaneously slid to the right to plug the empty spaces. As they filed past the cages to gather the machine guns for the march back to Harmony Church, J.W. grabbed the barrel thrust at him by Poliak. The still-steaming metal melted the fat in J.W.'s palm, and the gun slipped through his grasp, clunking noisily to the ground. It was not until push-up number twelve that the stink of burned flesh blended with the cordite and phosphorous to turn his stomach. He dipped his hand in a puddle of wet clay every few minutes on the march back to Harmony Church.

In the morning, he presented a silver-dollar-sized palmar blister to the staff sergeant and asked to be taken to sick call, or at least have his hand dipped in scotch, his mother's treatment of choice for childhood burns.

Hartack patted his own pockets and smirked, "Sorry, fresh out of Johnny Walker." The sergeant bent forward and unsheathed a hunting knife from his boot so quickly, J.W. did not have time to pull his hand away. Hartack lunged and unroofed the lesion in a fluid stroke. Amused at his surgical prowess, he ambled off, leaving the wound oozing a translucent, straw-colored, Knox gelatin. J.W. removed his ragged tee shirt, bound his hand, found the trunk of a great oak, dropped to his butt, and lit a cigarette.

Through the curls of smoke, J.W. thought about the gaps in formation, the ghosts of troops, those riding deuce-and-a-halves back to Main Post. He wondered which of those men would die in Viet Nam first, and if the Rangers would speak those names in whispers over the coming months.

Had LaVoy learned anything? Would he again dive into withering fire as he had on his first patrol? Would he run in front of his men to martyr himself in the pursuit of honor? Or had it been approval? Would he act even more foolishly to wipe the stain of failure in Georgia from the cosmic tally? Would this disappointment be added to the list he reckoned every night waiting for sleep? Would it set the cascade of failures in stone, add yet another shovelful of coal to the steam engine that could stop only when his name was inscribed on a dusty plaque in a city park in Iowa or Nevada? Would LaVoy, like most tragedies, be forgotten in four days? J.W. could not have predicted that, by the end of his first week in Ranger School, he would not be able to recall LaVoy's face.

Cowsen announced, "Rangers, the nucleus has been chosen. You are the privileged who have been chosen to carry the torch." J.W. exhaled, believing those who had been through the final test of courage were certain of graduating. They had paid the life insurance premium. As of that moment, they would be treated as humanely as the rumors had promised. It was with that sense of relief he and most of his fellows slopped through the obstacle course one last time.

On the run back to Harmony Church after midnight, however, Cowsen prodded them, "Gentlemen, close in those gaps, and do it on the double." Rumor had it two more Rangers had refused to drop into the underground tubes.

They were released to sleep. It was 2:30 A.M. Most collapsed directly onto the metal springs, using the thin, furled army mattresses as pillows. Fricker bitched that all the blankets had been stolen and got off his rack announcing, "I gotta find Sergeant Hartack and tell him what's goin' on."

Branch shook his head. "You outta' your gotdamn mind, Ranger? Shut up and go to sleep."

A few troops scratched razors over their faces, donned fatigues on which the mud had dried, then covered up against the cold, using as blankets the muddy, wet uniforms they had just stripped off. Most fell asleep dreaming repeatedly, as did J.W., of machine gun fire, and of traipsing, hour after black hour, on crunching twigs and snapping twigs on the hills, then velvety, soft twigs in the valleys near streams.

Fricker jumped up several times lamenting that all he could hear was the crush of deadfall and the slosh of creeks and rivers. He hollered in his sleep, "Every time I put my foot in the water, it freezes right there. Man, I can't move. Pull me out. Pull me out!"

J.W. ignored Fricker's raving and finally dozed off. But he, too, suffered a hallucination, witnessing himself outside his body, the last thread of his fatigue pants rotting away before his eyes. Underwear that chafed the first couple of days had been abandoned, and there was nothing left between his manhood and the seventeen-degree outback. His business hung out in the breeze like a limp, moth-eaten windsock at an abandoned landing strip. J.W. was exposed to the world, and while, a week before, public nudity would have constituted the lowest point of his life, that night, he could not muster the energy to care.

Long before the sun was scheduled to crest the eastern sky, J.W. woke fitfully from the fifty-first repeat of the nightmare. He shook his head wildly. He was incredulous and smiled inwardly. "Oh, my God! I believed it. All of it. It all seemed so real." He recognized the bedroom at Fort Meade; he could feel Krista's warmth as she slept beside him. There had been no Ranger School, no Georgia, just a hallucination sculpted by the orders to commando school that sat on his bedside table. His relief was muted, however, by the mental image of those orders, and he allowed himself a moment before reading them again, savoring the seconds left in bed, nuzzling his wife tenderly, saddened that he would miss her more than he could imagine. Then he laughed at himself for having suffered over so foolish a dream. Nothing, not even Ranger School, could be that dreadful.

He allowed his eyes to open slowly. For a moment, he was mildly confused. The befuddlement, though, deteriorated into mild agitation as he began to wonder if Krista was next to him at all. Then he awoke sufficiently to believe she was there, and he clutched at his mildewed mattress, holding her tightly. But when his consciousness sharpened, he began to wonder why Krista smelled so bad. Then, with a few more seconds awake, the confusion deepened. He wasn't really holding anyone or anything. Finally, he looked to his side to see his quasi-Ranger buddy Branch. They were trudging a few yards apart, crossing a frozen clearing enveloped by the winter forest.

J.W. reassured himself this was just the latest iteration of the same dream, though it struck him as odd that he could know Branch if he hadn't yet met him. Then it all became clear. "That's it," he whispered to himself, "I flunked out. Fuck me. Yeah, but so what? I'm warm."

The next sensation, though, was not as agreeable. It was the dread that he was about to board a plane for Viet Nam, assigned to die at the hands of the First Infantry Division.

He again reached out to snuggle Krista, to savor her warmth, but no matter how far he stretched his arms, he grasped only a thin, cold emptiness. His face began to burn, as if stung by ice crystals whipping in the wind. He rubbed his eyes open with frozen hands, detecting not a petite wife under the comforter, but an olive drab mass slumped at the base of a tree and a staff sergeant standing above the glob jeering, "Ranger Morelotto, I'm going to count to three, and if you're not off your fat ass, you're going to carry me for the rest of this mission."

Slowly, J.W.'s knotted consciousness uncoiled. Try as he might to pretend, he wasn't at home with his wife, and even the image of having returned to the barracks the night before had been an illusion. He perused the crop of new bruises and lacerations on his face, arms, and the family jewels, a badge of the twenty-kilometer mission through which he had apparently slept—on his feet. He massaged his eyes for another try at orienting himself, but there was only a patch of dark land in the distance toward which Cowsen and the staff sergeant broke ranks to run. J.W.

focused on the terrain, then exhaled in relief—they were back at the bleachers.

J.W. groaned as he eyed the pile in front of Cowsen's lectern: a two-foot heap of decades-old, olive drab C-rations, steel cans containing vitamin-laced Crackers and Cheese; vitamin-laced, soggy Fruit Cocktail; vitamin-laced, powder-dry, silver-dollar-sized discs of dark Chocolate; and a dozen main course selections fabricated from the permutations to which the Army had subjected beef and pork. J.W.'s internal alarm pounded most vociferously when he spied the Fruit Cake, an unsavory, desiccated dessert of barely-baked, vitamin-laced, lard-laced flour, poured tightly into a tin only slightly larger and harder to open than a can of tuna. To its credit, Fruit Cake was packed with "fruit," though, on closer inspection that turned out to be sugarized, crystallized, bits of orange peel, cherries, stems, and, to his dental officer's delight, the occasional pit. It, too, was flavored with the FDA-times-one-hundred minimum daily requirement of vitamin powder, which imparted all the appeal of the ABDEC vitamin drops his parents had squirted down his throat as a child. The "cake" moiety of this offering was a greasy, yet sub-Saharan dry, light brown, carbohydrate preparation. J.W. considered kneeling in prayer to beg for any alternative to the canned cake.

He stood to Branch's left, eavesdropping on a conversation in which the little man confided to Bearchild that he loved the Ham and Limas so much, he had made many a one-sided trade over the years to procure a single can.

"I got no regrets," he smiled to Bearchild. "They remind me of the ham hocks and peas my mama made when I was a kid. That's the best food on Earth, my man. Ever tried 'em?" Bearchild shook his head no. "I'm sorry," Branch said sadly. "You really missed out comin' up, man."

J.W. was confused. He thought back to the discarded C-Ration cans he had kicked around back in the ravine that first morning, and how all had been licked clean by his Ranger ancestors, all except the Ham and Muthas. He had tried them in the past, and no matter how hungry he'd been, after the first spoonful, there was no choice other than spit out the crunchy, gritty, overgrown, yellow beans and lard-streaked ham. He recalled having flung the can so hard, it landed in the next time zone. As he tucked away Branch's admission, J.W. laughed at the notion that Ham and Limas could ever bring anyone anything but cramps and gas.

The company waited impatiently for the moment Cowsen turned them loose on the pile of Cs, but first there came another lecture on knot tying, and then a treatise on foot care and the importance of changing socks in the field.

When Cowsen finally delivered the command to approach the aggregation of cans, he warned, "You will do it in an orderly fashion, one by one. You are soldiers, not savages. Secure only the C-ration closest to you. Do not pick through the pile, Rangers, this ain't the A&P."

Though the solemn order had issued from the lips of God, Class 68-B sprinted to the collection en masse, digging voraciously for favorites. J.W. swung elbows until he unearthed a tin of Franks and Beans. He slithered off to the back of the bleachers, where he hunkered on the edge of the gulch, cradling his ambrosia, holding it against his belly, guarded by extended elbows.

Cowsen waited patiently until the anarchy abated and each man had claimed a can and a patch of land. He watched without expression as the Rangers protected their selection, chests puffed, searching left and right for would-be robbers. He tsooked and shook his head. "Fall in." And the plasma of shredded olive drab groused into three abbreviated platoons.

Cowsen spoke softly. "Gentlemen, in Ranger School, we must learn to obey orders. Now turn to the man on your left. Trade your can with him. No, Ranger Branch, not to the man on your right, to the man on your left." Branch hesitated, staring longingly at the black label stenciled on the top of his meal. He bit his lower lip and stuck his arm a quarter-way out to Weathersby.

J.W.'s jaw slackened as he snatched the small man's dinner. He smiled broadly and licked his lips. "I live for this stuff."

Branch followed a pace to Weathersby's rear. He held the Franks and Beans behind his back. "Okay, man, two cigarettes and the tube steak for the Ham and Muthas. And I'm not goin' any higher."

"Four smokes and half your chocolate," J.W. countered.

Branch harrumphed. "Honkey always think he can stick it to the black man, don't he? No deal. Anyway, white man hates that lima bean shit."

Branch slowly spread his thighs apart and began to put the can of Franks and Beans in his crotch to warm it, but J.W. grabbed his arm and sputtered, "Okay, two smokes and a little bite of the chocolate. Deal?"

Though Branch scowled, a covenant was struck in the frigid dawn of the Georgia-Alabama border, and they sat anxiously awaiting the provision of tiny P-38 can openers. When none were proffered, Fricker went to the first sergeant and asked, "How we supposed to open these?"

Cowsen answered, "Not time to eat yet, Ranger. We'll tell you when. Got a class to do first." He pointed to the bleachers, and the company sulked to their customary seats.

Chalkboards and props were set in front of the benches. On the center table were several Claymore mines, anti-personnel devices the size and shape of a thin bible, but curved like a crescent moon. Inside the weapon were sheets of C-4 plastique, a deadly explosive impregnated with hundreds of miniature steel ball bearings. The mine was designed to explode to the convex, the outer-rounded side of the crescent, and shower a broad slice of the enemy with the metal pellets. Imprinted on that outside surface in bas-relief was the alert, THIS SIDE TOWARD ENEMY. That admonition had been carefully crafted to warn remind commanders that, as opposed to machine guns, Claymores were to be handled only by troops with an IQ of sixty-five or better, and those who were able to read English.

One of the main uses for the Claymore was to protect small groups of soldiers who had been sent out of base camps to act as listening posts, early warning patrols for the main body of soldiers. The forward troops set their Claymores many yards in front of them, then ran an electrical wire back to where they had dug in. When they heard the enemy closing, the soldiers flipped off the safety on their end of the wire, pushed hard on the crank of the detonator, and the mine went off instantly, propelling supersonic steel balls outward over a pie-slice of territory.

In Viet Nam, however, the entreaty about which side faced in and which out was lost on the young Vietnamese soldiers who had been armed with donated U.S. ordinance. The Saigon troops did not bother to have those instructions translated, for they had not an inkling the raised words were life-saving directives. Each time an indigenous Vietnamese soldier failed his language lesson, that troop didn't live long enough to warn his compatriots that a modicum of English was required to survive on the battlefield. Since Vietnamese listening posts were generally far from the main contingent of soldiers, the bodies of the victims who had placed them backwards were not discovered until the next morning.

There had been no impetus to provide the Vietnamese soldiers with reading comprehension curricula, for it was assumed that the mines had been set out correctly, but that enemy Viet Cong sappers had infiltrated stealthily, crawled up to the mines, turned them around, then jumped up and down making a ruckus. The South Vietnamese soldiers, delighted in the coming easy kill, confidently flipped the switches on their detonators and instantaneously blew themselves to their final reward. Toward the end of the war, when the truth was discovered, a sticker printed in Vietnamese was placed on the Claymores.

J.W. was a past master with these mines, having learned several tricks one could perform with the weapons from combat veteran troops at his first duty station, the 6th Cav, in Maryland. Most of the men in the platoon he commanded were Viet Nam returnees who had spirited ordinance and other matériel home from the war. They'd shipped starlight scopes, pistols, even AK-47 enemy assault rifles, mostly through the mail over many months, piece by piece, to avoid detection. One of J.W.'s problem children at the Cav, Private Morris Exton, had eschewed the mails and simply carted two Claymores home in his luggage. His plan was to scoop out globs of the clay-like C-4 explosive from inside the mine and use them for barracks pranks. He had dreams of burning the boots of his fellow soldiers, for if the Cav was anything like basic training and his unit in Viet Nam, the troopers would be teasing him mercilessly within two days of his return to the States. He was, however, soon relieved of his plunder by the very troops against whom he sought revenge. These men found it enjoyable to flush fiercely burning lumps of C-4 down the 6th Cav's barracks toilets, then blame the deep-thud explosion, and its attendant flood, on Exton, who was repeatedly awarded Article 32s, penalized with barracks arrest and toilet cleaning duties for weeks at a time.

J.W. looked about the area around the bleachers. The cadres had abandoned them, so some of the men used their bayonets to pry open C-ration cans and wolf down the first meal in days; three abbreviated platoons of America's exceptional hunched over protectively, using bare fingers to scoop clotted food from cans.

Anxious to capitalize on his prior Claymore training, J.W. pilfered one of the mines from the display table and carried it under his fatigue jacket to a corner behind the bleachers. He pried off the back with his bayonet, scooped out a golf ball of the putty-like C-4, snapped the cover back into position, and gave the mine to Fricker, warning, "Hey, man, put this sucker back on the table before the cadres miss it. Then I got a surprise for you."

Fricker was nervous and backed away quickly when J.W. touched a match and brought it toward the glob of C-4. "Calm down, man. This shit only explodes when it's detonated with a blasting cap. But it burns like a mother if you touch it with a match. Watch this." Fricker came back cautiously, and the two heated their meals over the white-glowing plastique until the olive drab paint on the outside of their cans vaporized. J.W. invited a few of the others to heat what was left of their meals. Branch was the last to finish. He invested particular care to extract the last film of the gravy with his index finger, finally sucking noisily at the micro-crescent of pot liquor remaining under the inside rim until he sliced his tongue. He cursed and threw the tin into the gulch.

Fricker shook his head, "Man, that thing was so clean, you coulda put it back on the shelf right now, refill it, and hand it to Class 68-C."

Branch had wrapped his fatigue jacket around the wound in his mouth, answering only with his eyes.

The demolitions class that followed dinner introduced the art of blowing bridges with C-4 explosive molded into the bottom of champagne bottles, setting road mines made of C-4, and fashioning detonating switches from old flashlight batteries, twigs, and the aluminum-covered paper at the top of cigarette packs. J.W.'s favorite seminar taught the destruction of a car by wrapping adhesive tape around the spoon of a hand grenade and then pulling the pin.

"Gentlemen," Sergeant Poliak, the machine gun maven, grinned, "drop that puppy into the gas tank of someone you care for, and after a couple of hours, the gasoline dissolves the tape, the spoon flips off, the grenade is activated, and four point six seconds later, the car explodes in flames. The most beautiful part of it is you, the saboteur, are miles from your deed. Rangers, you can't beat that with a stick!"

The students listened excitedly through that discourse, for they had been promised an opportunity to actually blow the mines and try the tricks they had been taught. But before the lab practical, they had to sleep through several more lectures, the last a dissertation on toilet protocol in the forest.

Cowsen, who had been sitting in the grandstands in rapt attention during the final lecture, emerged from the bleachers and began handing out the various weapons, presenting J.W. with the very Claymore from which he and Fricker had made the withdrawal.

"First Sergeant, I kinda wanted the champagne bottle with the C-4 in it, if that's okay."

Cowsen smiled, took the Claymore out of J.W.'s left hand and slapped it into J.W.'s right more forcefully, ordering, "Stop flapping your lips, or I'll glue 'em to the outside of this here device. Now set the mine out there like you been taught in class today, and get behind the bunker. I'll tell you when to blow it, Ranger."

J.W. positioned the device, struggling to remember which end suffered the deficit of C-4. Balancing it on the tip of his index finger, he surmised it was on the left and aimed the apparatus far to the right, then ran the wire back to the musty, mossy, Second World War bunker in which he and his compatriots would hide during the blast. As he jumped behind the barricade, his boot lightly brushed the crumbling concrete. The chunk of the cement that toppled off dropped on Cowsen's spit-shined boot. During the push-ups, J.W. occupied his mind by picturing the diseased Claymore blasting backwards and killing him. That thought slowed his calisthenics to a snail's pace until Cowsen kicked his arms out from under him.

With J.W. lying in the prone position in the fragile trench, Cowsen blared, "Do it, Ranger!" And J.W. sucked in a deep breath, squeezed his eyes shut, and cranked the switch. The mine ruptured with a blistering crack, not the thud of the other Claymores. It flipped hard right and broke in half. One piece spun into the gulch like a fluttering maple seed, inciting a cheer of appreciation from his compatriots; the other fragment arced over the bunkers, landing on Cowsen's lectern.

"What artistry," Gillette complimented.

Cowsen grunted in disgust, "Your time's a comin', Ranger. The rope's a gittin' shorter and shorter. And you, Ranger Gillette, drop and give me twenty-five-and-one."

⚔

They sat in class for several more hours, reawakened with pauses for push-ups each time a critical mass of Rangers' heads bobbed. At midnight, they were given a break, but Sergeant Poliak warned, "No lights, no matches, no flashlights."

Fricker raised his hand and asked, "How you supposed to smoke?"

Poliak did not answer but commenced the next lecture early, explaining, "Rangers, it requires at least twenty minutes of pitch black to hone your night vision. Do not stare directly at what you want see at night. Look at a point twenty degrees to the side of target. That allows the light to fall on the optic nerve, the most sensitive part of the retina."

Gillette raised his hand. Poliak sighed, "What is it, Ranger?"

"Sergeant, it is the fovea of the eye, not the optic nerve, that's receptive to weak light."

While Gillette was working off his sentence, J.W. tried the stratagem of looking to the side. He raised his hand. "Sergeant Poliak, thank you. That's the first useful bit of information I got out of this school."

Sergeant Poliak did not wait for J.W. to finish his penance before communicating via a handheld PRC-25 radio-transmitter with a counterpart several miles deeper in the woods. Poliak spoke into the microphone, "Start display number one."

In the distance, a tiny flicker of yellow appeared. An instant later it became bright white, faded, reappeared, faded again, flashed, then disappeared.

"Rangers, was there ample time for a sniper to aim his weapon and fire at least once at the distant light?"

When most nodded, Poliak went on. "What did you see, Rangers?" After a long pause, Bearchild called out from the balcony, "A match."

"Gentlemen, that is exactly right. Three soldiers lit cigarettes. Anyone heard the expression, 'three on a match is bad luck?'"

The next demonstration was an auditory cue, the clanking of metal far in the distance and the barely perceptible sound of grinding earth. Again, only Bearchild was sufficiently awake to answer. "Tailgate. Deuce-and-a-half off-loading infantry."

"How do you know it's infantry, Ranger?"

Bearchild thought for a moment. "One truck, ten men crunching gravel, one squad. Who else is dumb enough to be out walking at night?"

Poliak laughed. The rest of the class turned to stare at Bearchild, including J.W., who had barely heard the sound of the clanking metal. Bearchild was holding to his ear a sheet of paper rolled into a cone as if an eighty-year-old listening to his wife.

Poliak droned on until the rain began, as it did late each evening, the showers a signal to the cadres to order the company further into the forest and away from shelter. They halted at another set of decrepit bleachers, taking seats in the drizzle. Poliak pointed to the northern sky. It glowed in green-blue, horizontal streaks, an eerie, monochromatic rainbow. The show went on for miles, fading finally into the blackness.

"The aurora borealis," J.W. called out.

"Nope, not even close." Poliak sighed and shook his head sadly.

"High tension electric lines," proffered Gillette, and Poliak nodded in concurrence until Gillette added, "ionizing the atmosphere. The field surrounding the flow of electrons of that potential..."

"Shut up, Ranger."

The company was sent marching in a quest for Gillette's lights, perceiving a static hum long before reaching the wires. Coming closer, they could make out droplets of rain roiling about the cables, glowing in foot-diameter shimmers of turquoise, more a disco than a soaking, forest night. The waves of static electricity made J.W. tense at first, then overtly jittery as they closed in on the high tension power lines. It left him with a chest as tight as if he had guzzled too much of the coffee he had not had in nearly a week.

The company was dismissed to barracks at 3 A.M. J.W. asked Branch if they were really indoors, or if it was just another dream. Branch stared at him irritably but dropped onto his bedsprings and fell asleep before offering up a homily.

J.W. tossed for a long time seeking warmth, thinking about food, and crying over the threat of going on. He wondered if his dad had suffered through as much during the nearly half-decade he'd spent in the Pacific war. During those endless, sleepless years without a break in combat, had he ever thought of quitting? "Pop, how did you do it? I'm what, five days into your routine, and sinking fast."

By 6 A.M., they had finished the second loop of the obstacle course, despite the promise they would not see it again.

Fricker laughed. "Easier than it was the first day, huh?"

Morelotto appended, "Yeah, but no goddamn cleaner."

At the mess hall, Cowsen spoke quietly. "Pull-up price now gone risen to thirty."

Gillette wagged his index finger above his head. "First Sergeant, that's an increasingly inverse, no, no, perverse, relationship between the number of pull-ups and the quality of the chow. Are aviscous mashed potatoes the entirety of the Army's gastronomic offering?"

Cowsen stared at him for a moment and answered cautiously, "That is correct, Ranger."

There was less supervision in the mess hall these days, and Rangers nicked the odd packet of sugar from the cadres' section. There were also containers of milk, though the latter lay untouched after Gillette lectured, "It doesn't require a degree in physiology to deduce that it's the protein, the casein in the milk, that curdles in the stomach. That's what causes the stitch in your side during the after-dinner run. I'm not touching a single dairy product for the rest of this course."

J.W. guessed he'd already shed fifteen pounds when he looked at himself in the mirror on the sixth morning. At first, he didn't recognize the gaunt, bruised face as he started to shave, but each stroke of his razor matched that of the apparition staring back at him, so he guessed he really was looking at himself.

After toiletries and sixty-five seconds in the mess hall, the cadres resurfaced and the company double-timed back to Victory Lake, the south end of the lagoon this time. They were to execute the "Slide for Life." Erected along the shore was a seventy-five-foot-high telephone pole capped with a flimsy, three-foot-square wooden platform. A slack, rusting, metal cable fell in an upside down parabola from it to a short telephone pole staked three hundred feet into the lake.

An E-7 took a place at the front of the company. "Rangers, welcome. Each of you will be issued a pulley welded to a handlebar. These are delicate devices, just like you. Treat them with respect, as you have been treated by the cadres. You will climb to the top of the pole and place the pulley over the cable, grab the handlebars, and on command, you will jump off the platform. Those of you who have taken seriously our efforts to teach you the art of hanging onto metal bars and ropes will survive the fall. As you slide down the cable, reaching sixty or seventy miles per hour, and maybe eighty, for those of you who are still fat..." Gillette's hand flew into the air. "What is it, Ranger?"

"Excuse me, Sergeant, but one's terminal velocity is not determined by his mass, unless you consider the extra air resit..."

"Okay, Ranger, let's do a scientific experiment. Let's see how many times you can move your ass up and down. Try fifty-and-one."

"No, Sergeant, I said, 'mass' not ass.

"And another fifty-and-one. May I continue?"

"Proceed."

The instructor opened his mouth but stopped to glower at Gillette. "And *another* fifty, wiseass. As I was saying, you will slide down the cable, reaching...ah, high speeds by the time you smash into the far pole. You will notice that surrounding the far pole are a few truck tires. They are old, gentlemen. The rubber has hardened to stone. I don't know, but there is a chance they might not protect you. Our suggestion to avoid death is to kip, that is, bring your legs up parallel to the water, and drop into the lake like an 'L'. It is recommended you do it before the end of the run. Land on your butt and skim to a stop before the pole. But, again, that's just a suggestion.

"Now, Rangers, there is a conundrum here. The sooner you let go, the farther you drop. The longer you hold on, the faster you hit the water, and the closer you get to the pole."

Fricker's hand flew up. He turned to Cowsen. "First Sergeant, what is your opinion?"

"Ask Ranger Gillette."

The staff sergeant continued. "Gentlemen, I draw your attention to the sheet of ice on the surface of Victory Lake."

In the deep recesses of J.W.'s mind there was an image of this very scene, though it wasn't until the cadres disappeared that he understood. But by the time he, along with several of the relatively alert Rangers, dove for the bleachers, the explosion had already heaved an ice-laden tidal wave of freezing slime onto the company. A round of derisive laughter came from the cadres as they exited their heated lean-to.

Poliak stationed himself at the base of the pole, his mission to dispense to each cadet a handlebar welded to a pulley. He called Morelotto front and center to climb first. "You dodged the bullet last night, Ranger, but your daddy ain't gettin' you outta' this one. Get your fat ass up there. And Professor, you're next."

One of the higher wooden rungs split under Morelotto's mass, and he fell a few feet. As his boot struck Gillette's head, he caught himself and started up again without a word of apology to Gillette, whose eyes were rolling in twitching circles.

On the platform, Morelotto stood wavering in the wind until the cadre on platform duty wrested the pulley out of his clenched fists and placed it over the wire. A millisecond later, the man booted the trembling cadet in the ass. Morelotto slid for less than thirty feet before he tried to kip. His legs, tree trunks wagging in the air, came up five or ten degrees, and with a contorted face, he let go of the handlebars and tumbled toward the lake sideways. Staff Sergeant Hartack spit into the water, climbed into the rescue dinghy, and howled, "Shit, that didn't take long."

As the big man tried to climb aboard, the boat nearly capsized. Gallons of the frozen water roiled in. Hartack's spit-shined boots were soaked, and his eyes blazed in fury. He shoved Morelotto away, though finally allowed him to hang onto the stern to be towed in like a whale carcass.

Gillette performed a perfect kip and adjusted his legs as if rudders, skimming gently to a stop near the shore. Fricker was next up. He asked for a hammer and nails to fix the rung Morelotto had broken. Poliak jammed a pulley into his hands.

J.W. was summoned for the ensuing turn. He complained, "Sergeant Poliak, the weld bead on my pulley is cracked. That's dangerous."

Poliak rolled his eyes. "Of course, Ranger. Safety first." The grizzled sergeant snatched the pulley from him, handed it to the Ranger behind J.W., then searched his cache of pulleys, chose another, held it up to the light for close inspection, nodded, and presented it to J.W.

Fricker, who had taken only two steps up the pole, examined his own device. He took a breath to speak, but J.W. cut him off. "Hey, Sergeant Poliak, this one's cracked worse than the first one."

Poliak snorted, "No one said the Rangers in the class before you knew a damn thing about how to weld." He pointed an index finger at Fricker and then a thumb toward the platform. When Fricker had ascended four more rungs, Poliak trained an index finger at J.W.'s face. "Get goin', Ranger. Let's see if you can *slide* forever."

At Morelotto's fractured rung, Fricker decided not to continue and took a step down. Poliak screamed up, "Three points of contact, Ranger." Fricker did as directed, letting go of his pulley, which clunked onto J.W.'s head. It bounced up nearly back to Fricker before tumbling to the ground. Fricker continued down, but when he caught sight of the blood squirting from J.W.'s scalp, he gagged, swung his leg roughly up one level, and caught it on a nail. Poliak jumped aside to avoid the gush of blood.

Poliak yelled up, "Weatherman, Frigger, both of you halt! Weatherman, hand Frigger your pulley. Frigger, take Weatherman's pulley and get your ass up the pole!"

Fricker called down, "Begging your pardon, Sergeant, but I want my old pulley back." He started back down. J.W. beat him to the ground.

Poliak took J.W.'s handlebar, gave it to the Ranger behind him, handed Fricker one from the box, then gave J.W. his original. When J.W.'s mouth opened, the sergeant growled, "Just fuckin' drive on, Ranger. Jesus Christ."

The cadre manning the platform swayed in cadence with the winter wind, seventy-five feet above the lake. He grabbed J.W. by the back of the pants. The gesture reassured Weathersby, and he relaxed a bit. The sergeant took the pulley from J.W., swung it over the metal cable, then directed, "Stand on the edge, Ranger. Don't worry, you'll be fine. I'm holdin' onto you. Now grab the bar. Lean forward just a little." As J.W.'s fingers touched the handlebars, the sergeant let go of his pants and nudged him in the back. When J.W. s toes dug in, the man kicked him in the butt.

At first, the drop reminded J.W. of airborne school, the wind hitting his face harder and harder as the seconds passed, but the snap he had come to cherish in jump school, the moment the life-maintaining canopy burst open and caught the air—that was tardy. He waited, but as he approached sixty miles per hour with the canopy yet to deploy, J.W.'s body weathervaned into the wind—backwards. Kick and twist as he might, he could not spin his body to face frontward.

Cowsen yelled, "Drop, Ranger!"

But J.W. did not hear the command, preoccupied with increasingly violent squirming and the mounting roar of air. Poliak joined in, "Drop, Ranger!"

Still, J.W. held fast to his pulley. It was not until Cowsen, Poliak, Fricker, Gillette, the staff sergeant, and Branch shrieked, "DROP, RANGER! NOW!!" that his hands opened.

Touchdown commenced with an impact on a patch of ice-soup. The shock was absorbed by his neck and back. He winged on like a skipping stone, still backwards, up onto the thickly-glazed portion of the lake. Contrary to Newtonian physics, he gathered speed instead of slowing. J.W. turned his head to look behind him, actually in front of him, to see his sliding expedition on a trajectory aimed directly at the anchor pole. He grabbed at the first safety tire, but his arm jerked painfully and he had to let go. He was sufficiently awake at this juncture to appreciate that the Slide for Life ended at the far pole, and so would his stay at the Ranger School.

At the second tire, instead of using his arms, he jammed his foot into the hole. When the tire came to the end of its fifty-foot hemp tether, J.W.'s back wrenched, sending an electrical charge zinging into his right leg. The rope held, though, and he was propelled into a sweeping arc, a demonstration of angular acceleration worthy of a high school physics experiment. The rope, taut with J.W.'s momentum, whipped him around the pole back toward the defrosted portion of Victory Lake. He timed the removal of his foot from the tire to allow his body to be slung into the water, where a weak sidestroke and the swells he had created moments before carried him toward shore.

There were shouts of appreciation from the gallery as he dragged himself to the water's edge. It gave him courage, though not enough to stifle his cries that he had really hurt his back. He crawled onto the sand and waited for the pain to wither. Hartack glanced over from the rescue boat, took a few oar strokes toward shore, but noting Branch was next up on the slide, tossed the smoke overboard, and rowed out in preparation.

J.W., still writhing, rolled on his side to watch Branch make his drop but lost him in the sun. In a few seconds, Branch dog paddled to shore and, after nearly tripping over J.W.'s hulk, hauled the moaning Weathersby to the bleachers as though dragging a fallen D-Day comrade from the beach at Normandy. He deposited Weathersby in front of the stands. "Hey, man, you're covered with dirt," he spat crossly, then climbed to the top level of the bleachers and lit a Chesterfield.

The longer J.W. lay on the ground, the tighter his back spasmed, and while the others had taken to milling about to keep warm, J.W. was relegated to generating heat by shivering. When not a single cadre commanded J.W. to get off his lazy ass and join the main contingent of Rangers for the voyage back to Harmony Church, he considered remaining there on the ice-hardened, sandy beach for the rest of his life. But as the cold burned more deeply into his chest, he thought better of dying on that spot, and dragged himself to his feet. He ran scrunched forward, limping in concert with the seizing pain that gripped his back. A mile on, he sensed a peculiar numbness growing along the outside of his calf. He called to Fricker, "Shit, man, I think I had a stroke."

Halfway through the run back to Harmony Church, J.W.'s mood lightened; the lack of feeling in his leg was a godsend. It radiated both up and down his leg, masking the pain. Two miles later, though, a deep soreness began. It soon deteriorated into an agonizing ache from his butt down the back of his thigh, past the knee, into mid-calf. As the throbbing bored deeper, he slowed and considered collapsing in a ditch along the dirt road and not fighting back when the staff sergeant pummeled him into brotherhood with Kenyon.

The lecture that afternoon scrutinized the art of hand-to-hand combat. Sergeant Poliak directed the company to close their eyes and imagine they were on a great, barren desert with nothing but fine sand for a thousand miles.

"Rangers, you come upon a sentry that you must neutralize. You cannot make a sound. You have no weapon. You possess only your toilet kit and your hands. What do you do? Well, I'll tell you what you do. You put a bar of soap in a long wool army sock, sneak up behind the son of a bitch, swing that sock hard, and pop the sucker in the head. Rangers, this is a hell of a bludgeon."

Fricker, who had been nodding off, awakened, suddenly alive with eyes gleaming, apparently interested in the subject of exotic weaponry. He began to raise his hand, but Branch pulled it down and hissed, "Keep your arm down. You hear me?"

Fricker, however, sat erect as Sergeant Poliak produced from a burlap bag a quiver of piano wire cut in two-foot lengths—thick, bass note strands and thin, silvery alto wire.

"Technique number two, Rangers, entails attaching a small stick to each end of a piece of tactically useful piano wire. Shoot, you can just wrap it around your hands if you don't have a stick. Now listen carefully. You sneak up behind the unsuspecting sentry, different guy than you hit in the head with the soap, 'cause you gave that one a brain hemorrhage. Now slip the wire around the new guy's neck, cross your hands rapidly and firmly over each other, and, voila, the enemy's head falls off right by your feet. It's as simple as that."

Noting Branch's head bob, Fricker's hand shot up and waved wildly until the instructor recognized him. "Hey, Sergeant, excuse me, but where do you get piano wire in the middle of the desert?"

"You're not in the desert anymore. Close your eyes and imagine you're in Carnegie Hall, ASSHOLE!" He calmed quickly, though, and added, "Now, please give me twenty-five-and-one for your failure to correctly assess your tactical location." When only Fricker assumed the position, the sergeant screamed, "All of you!"

As the men retook their feet, Poliak asked, "Who wants to volunteer to try the next technique, the stranglehold, on Ranger Frigger?" Those who hadn't fallen back to sleep raised their hands greedily, and J.W. considered joining them, but he thought back to his pop's admonition, "Never volunteer."

That stemmed from somewhere in the Pacific, when one of J.W.'s old man's sergeants asked if anyone in formation knew anything about music. The elder Weathersby raised his hand, as did the man next to him. The two spent the steaming, tropical afternoon struggling with a grand piano, hoisting it to the second floor of the Officers' Club.

J.W.'s hand was the only one that did not shoot up in petition to harm Fricker, so Sergeant Poliak surveyed the audience and pulled J.W. by the ear to the podium. "Strangle Forever," he snapped, "come with me front and center."

Fricker was forced to face the class while Poliak dragged J.W. back five steps, had him skulk up from behind, throw his forearm across the front of Fricker's neck, and pull it powerfully with his other hand.

Sergeant Hartack, who had been standing quietly on the side of the bleachers sipping hot coffee, strolled over. "Harder, Ranger," and then, "Harder, I said!" Fricker's face took on a peculiar blue cast, and his hands and arms vibrated in a violent tremor.

Poliak commanded, "That's enough, Ranger," and Fricker was released to hobble like a drunken sailor back to the bleachers and drop noisily onto the planks next to Branch.

Branch shot awake. "Quit movin' around, fool. I'm tryin' to pay attention."

Convinced by the terrified expression on Fricker's face, few of the Rangers felt the need to test the maneuver. Nevertheless, Poliak insisted, "Each Ranger, you will perfect the stranglehold on the troop to your front, and you will do it until you get it right."

Though being violently suffocated was terrifying and painful, and left the Rangers quietly anxious and not in the frame of mind for further instruction, that was not to be final chapter in the hand to hand combat curriculum. The company still had to polish their skills in the art and science of the eye stab, the cardiac death punch, and the *pièce de résistance*, the scrotal grip.

When the troops began to execute the maneuvers with curtailed gusto, Hartack groused, "You Rangers are going to do push-ups until one of you gets hurt from these skills."

J.W.'s mind, addled from the stranglehold Morelotto had put on him, drifted for seconds at a time to the fantasy of a ten-minute nap, and once, fleetingly, to the co-ed's breasts he'd caressed at Sterling College on his first night as a freshman. His thoughts, however, always returned to the image of a peanut butter and jelly sandwich.

As the sun crept below the hills west of the bleachers, class was adjourned, and the Rangers sailed into formation for the march home. Split into platoons, the men of the three units eagerly stretched in preparation for the run along the pockmarked roads back to Harmony Church. J.W. smiled, calculating that by cleaving the company into smaller units, the cadres could move the men to the mess hall far more efficiently. Surely also on the minds of First Sergeant Cowsen and his lot were the nuggets of hail blowing with snowballing intensity into the faces of eighty starving men, not one of whom was dressed for a winter storm—barely one was dressed for the beach.

The men moved with an uncommon willingness until Cowsen directed the three units to disperse onto diverse trajectories toward the woods. The shroud of grey that cloaked the men spun into abject black as Cowsen tapped a Ranger with the whittled stick. J.W. began to quiver, sick that one more mission would intercede before they chowed down. Then he understood—the exercise was simply an opportunity to train student leaders in the art of delivering starving troops to base camp from various bearings, inculcating the skill to read a compass from any angle. But the new platoon leader was handed a thick manila envelope, and Cowsen corralled several Rangers, herding them to the new leader's circle to accept tactical duties.

J.W. dropped into a squat in a tree-covered thicket, one with only a trace of the deepening snow. He would sleep for the precious minutes the new commander squandered orienting himself. After a few seconds, however, J.W. believed he was again dreaming, for the vicious chill was blunted by a waft of humid warmth radiating from the earth. When the column began to move again, he realized he had made a major discovery: by hunkering within the heavy stands of hickory and oak, one could harvest the forest's entropy, the dark energy saved from the day's sun and given back, though sparingly, to its nocturnal creatures.

The company moved so slowly, the troops at the rear fell into trudging sleep, marching on in body only. Though the soldiers generated as much noise as did the roiling winter thunderstorms, it wasn't long before Morelotto drifted off from the main body, forcing the platoon to a halt.

The new leader snapped at Morelotto's buddy, Fricker. "We gotta' find the asshole, or we don't get credit for the mission. Sergeant Cowsen told me that if you whistle like the whip-poor-will, and keep doing it, he'll get the idea and find his way back. He says it works."

Indeed, the power of the Ranger Bird's song revealed itself as Morelotto stumbled back into to the fold, dripping from head to toe, saturated with pond scum in various stages of freezing solidification.

"Well, look who's here," Branch sneered caustically. "Think I'll write a book—*Look What the Ranger Bird Dragged In.*"

On the other hand, the budding guerrillas were as yet un-schooled in the pitfalls of resorting to remedies so easily applied, particularly ones offered up by the cadres. It was not long before, as every class before them had eventually discovered, when the real whip-poor-will cried, and there were countless numbers of them in the forest, Ranger dreamers floated off from the main body. Some went missing for whole missions.

Hours passed that night with J.W. so deeply comatose, he remembered nothing of the trek except the hunger and cold. He had again been dreaming of home, of warmth, of forgiveness for the transgressions that had brought him there, a wickedness he must have perpetrated, but where and when he could no longer remember. Now and again, the dream of being home filled him with cheer, but not enough to ease the throbbing in his back and every other muscle and joint he exploited trekking the Georgia outback.

After midnight, when the column halted, he accordioned into Bearchild, slipped, and fell backwards onto the frozen earth. His back cracked, and an electric pain flashed again into his right leg as it had at Victory Lake. He tried to regain his feet, but could not rise to his knees, and finally collapsed, conceding there was no way to go on, and no way to stop the agonizing numbness in his leg. He rolled about, seeking a position that did not hurt, and in those gyrations, his frigid, nearly numb fingers came to rest on a slick, gelatinous substance coating the ground. As J.W.'s hands swept across the disagreeable wad, he thought perhaps he was touching the carcass of a poisonous animal planted by the cadres, another ploy to frighten a Ranger into bawling, "I quit this shit!"

He brought a bit of the goo to his nose. The mass was not edible, the scent so rank, J.W.'s hands retracted to his side in violent reflex, fear and disgust peaking along with the pain in his back. He rolled off the creature's remains, sliding along the viscous surface, then sprang to his feet and groaned in a bloodcurdling whisper, "It's a dead body, man! I think it's human."

Bearchild calmly jiggled the switch on his flashlight, shining it at J.W.'s feet. All that glowed in the weak light was a flattened carton, a gigantic cardboard box with "REFRIGERATOR" stamped in moldy, black letters. J.W. took a deep breath, squatted, and touched it again, now smiling bravely, accepting that he had been the victim of a tactile mirage. Though he wondered for an instant how it had found its way to that spot on Earth, to rot dozens of miles from the nearest refrigerator, a different train of thought sparked suddenly, and he lifted the box out of the snow. Though mildewed pieces broke away, he discovered solid, only damp patches toward the center, and he folded those tightly into a packet the size of a grocery bag, one he soon hallucinated was filled with bacon and eggs and bread and cookies and coffee. In the seconds before the platoon moved on, he finalized his decision to purloin the load, jamming it inside his rucksack, hiding it from the cadres. Despite the added weight, he marched on, a man reborn, driven by the machinations of Ranger-Think, a particularly intense variety of mental disease that allowed the human mind to transform a sheet of putrefied paper into a trinket of salvation.

As he trudged forward, it became clear what had happened when he'd slipped in the snow. His foot had struck a genie's urn, which produced the cardboard, an offering, a special gift laced with the power to provide warmth and, somehow he was sure, something to eat. For several hours, this notion deepened, and he drove on through the forest with assurance, relishing the sweet vision of the pleasure he would soon savor.

That moment came at 4 A.M. when Cowsen's magic stick waved, inaugurating a new student commander. The respite allowed J.W. time to unfold the gummy cardboard and hack it into soggy sheets with his bayonet, a piece to line the ground, the foundation, a segment to serve as a blanket, and, finally, a scrap for a roof over his head. Bearchild watched longingly until J.W. handed him the leftovers. Bearchild nodded subtly.

The two placed their ponchos on the freezing loam then unfurled a rasher of cardboard on top of the rubber sheet. They dropped onto the compressed paper, pulled the second sheet of cardboard over their bodies, and flipped the other half of their ponchos over themselves. Wrapped in adjoining cocoons, they slowly drifted off, escaping reality, sealed away from even the slightest invasion of the frigid air. When word filtered back, seven minutes later, to prepare to move out for the next sector of the forest, J.W. awoke dripping with sweat. Despite that disappointment, for a brief moment, the first in many days, all was right with his world— J.W. Weathersby was fat and delicious. He ignored that his fatigues were already icing up from the perspiration he'd generated in the hovel, his tattered pants encrusted in a layer of frost as thick as on a Piper Cub's wing plying the Alaskan peaks. He whispered to Bearchild, "What else could a man want?"

As the march resumed, however, their nirvana vaporized as the hunger, fatigue, and arctic cold reproclaimed themselves, reminding them that pleasure in this universe was ephemeral at best. Bearchild's fatigues were also solid with sweat-ice, but his expression did not change, and he toted his slice of salvation for hours, finally handing it back to J.W. with a mumbled, "Give it to Branch."

The next time their boots stopped shuffling, J.W. presented one cardboard sliver to Branch and the other to Fricker, both of whom dropped to the earth to fashion complex nests. Fricker positioned and repositioned his packaging until word filtered back to push off immediately for an assault. There had been no time for either Ranger to enjoy a moment of the gift, nor an opportunity to repack the bequest. In the fog of battle, both Fricker and Branch had relinquished their endowment back to the forest, where all grants in Ranger School were destined to revert. At dawn, when they confessed the loss to J.W., he ran back along the trail but found not a shred of the genie's benefaction.

J.W. yelled at Fricker, "Where the hell are your priorities, man? It'll be a bitter, damn day before I give you anything again."

Fricker thought a long time before he responded. "Yeah, well, maybe somebody from 68-C or D'll find 'em. It just goes on, man, the circle of the cosmos."

"Cosmos, my red, rosy ass," J.W. raged.

As the sun's cold rays peeked over the hills, the company converged on Harmony Church for two circuits through the obstacle course. No effort was expended to avoid the mud, for that sped the chore. Breakfast was fast, but Gillette mused, "Adequate. They've provided sufficient calories to cover the energy expenditure. And don't forget, our stomachs have atrophied to the size of walnuts, like the brains of Tyrannosaurus Rex."

"You mean like the brains of the cadres," Morelotto added more loudly than he had intended, bringing upon the company fifty-plus-one.

After a few inhalations of mashed potatoes and greasy toast, Fricker left the mess hall complaining, "Man, I feel bloated, like all of a sudden, I'm fat."

Branch shook his head. "Ninety seconds ago, you were bitching about days in the field without food, eighty miles on forced march. Now you're complaining about bein' fat? Ranger, your head is screwed on wrong. Just shut up and enjoy it."

Cowsen, having overheard the conversation, obliged Fricker and announced, "Rangers, Ranger Fricker here's feelin' fat. Rangers are lean and hard, not fat. So we will move at the double-time back to the bleachers, where we will have a special lecture.

This will be your last formal class in Ranger School, Rangers." A cheer rose. Cowsen's face drooped in melancholy. "Rangers, this instruction is for you. It is priceless. You should be screaming for more." He brightened. "Your misplaced happiness is going to be expensive—that's fifty-and-one. Get started."

At the bleachers, First Sergeant Cowsen introduced the speaker, a captain in a Class A uniform, tie perfectly wound, low quarter shoes as bright as patent leather. He stood proudly and announced, "This morning, we will discuss the menace of Communism. When we are done, you will understand more about the peril of the Eastern World than you ever knew there was to learn.

"Gentlemen, I just stepped off a plane from Viet Nam. You are getting the latest information available. You will be as up-to-date as President Johnson. You have been gathered here to become enlightened, to understand why you must learn to hate the Communists. I cannot begin to tell you about the atrocities the Viet Cong have committed against our soldiers, to say nothing of the terror they have unleashed on their own people. Their own people!"

The troops listened in rapt attention, waiting to be told of the honor and fearlessness with which the American soldiers had comported themselves, and then the inventory of horrors perpetrated by the Viet Cong, that poor country's Communist Party. The captain spoke of the "insanity of Communism," and "the need to stop the Asian madmen before they invade Long Island."

Branch leaned over to Bearchild and whispered, "Hey, man, I know that combat patch on his arm. This dude's a political advisor, some kinda information officer. Betcha he never left Saigon. He's a FAC."

Weathersby grunted under his breath, "FAC? Hey, man, that's a forward air controller. Don't laugh, that's dangerous duty."

"No, man, that's 'fuckin' armchair commando.'"

As the captain preached passionately, and his young audience hung on every word, Branch rolled his eyes in boredom and turned back to Bearchild. "That Bronze Star he's got don't even have a V Device for valor. All that thing is is a medal for showin' up in Viet Nam and not gettin' caught in a whorehouse. He ain't no hero. Man, that's just the officer's good conduct medal. Everybody gets one. This guy don't know diddly shit 'bout combat."

As Branch's disgust deepened, some of the troops behind him "shushed" the little man, and for a few minutes it worked, but every time the lecturer bristled with contempt for the Vietnamese Communists, Branch howled in a whisper, "Cut the shit, man."

Exactly fifty-nine minutes after he began, the captain, with the skin of his shaven head furrowed in loathing, raised his voice and declared, "These Communists would have you believe their system is better, fairer. I'm here to tell you it isn't. It is a system of slavery, and we all know it!"

The captain nodded to Cowsen, and there was a bit of polite applause from the cadres who had been briefed those would be his final words. The Rangers, however, were silent, waiting expectantly for a catalog of the ordeals this man had undergone, a detailing of the abuses he had suffered along with the people of Viet Nam. Where was the discussion of the penetratingly corrupt political system of South Viet Nam for which the children before him would soon die to preserve? Was the man going to talk about the ineffectual, abusive Vietnamese military, for which billions in taxes were being sucked out of their parents' pockets each year? But there was nothing more. The man was done. He tiptoed through the mud, took a seat in the rear of a military limousine, and was whisked off toward the mess hall.

Cowsen picked himself up smartly and ordered the company into ranks. "Rangers, there should now be reason enough to understand why we must be willing to sacrifice." There was a flourish to his commands. He was guiding the privileged few with whom the secret had finally been shared.

As they double-timed toward the barracks, J.W. shook his head and asked Fricker, "What did that dude say? I musta missed it."

Cowsen, from fifty yards behind, stopped calling cadence and shouted, "It means, Rangers, things are for real from now on."

For days on end, they slogged the frosted mud. J.W. woke occasionally from his walking sleep, hunger always his first thought, food and the never-ending steps weaving incessantly through his consciousness. At dawn each morning, with Sylvester cresting the eastern hills, there remained only the memory of constant movement, and J.W. sought to record that which he could remember, sneaking off behind a tree to scratch on the back page of his Ranger Manual images of the past night. Staff Sergeant Hartack, however, caught him one early dawn and snatched the book away, burning the paper to heat water for coffee.

Alone with his thoughts for days and nights, there were hours when J.W. began to hear dark, reverberant voices calling him, scolding him. The dissonant chant thundered in his ears as he marched and climbed and crawled the thickets and hills. With each sunset, the unintelligible declarations became louder. He cupped his ears against what had decayed into shrieking, then tried ear plugs made from the scraps of olive drab wool he'd torn with his teeth from his socks. Some nights he tried to sing in a violent bravado above them, but the voices quavered unceasingly.

There were occasional breaks in the marching, and on one slog after nearly three weeks of training, J.W. and Fricker rested against a tree at four in the morning. They smoked and conjectured as to why the company had been allowed to rest. Though the woods were crawling with cadres, J.W. took the opportunity to recount to Fricker the story he'd overheard the night before. It was a sordid tale passed down from class to class.

"Didn't you hear it? Word is, Captain Vock's mother's a nut case." J.W. related with sarcastic laughter, "She's been in like a dozen loony bins, and still she comes out here to Benning when she's out on parole. Guess she stands around and yells at him."

Fricker smiled and asked, "You mean out here in the woods?"

"I don't know if she comes in the woods, man, but she's supposed to have been in Harmony Church. That's what I heard, and then, get this shit, she's like turned him into a homosex...," but J.W. heard rustling behind them, and sensing the presence of cadres, he snapped his lips shut. He surreptitiously turned to his rear, and behind a tree, not eight feet away, a gaggle of instructors stood silently in the darkness peering at him. J.W. could not be sure, but he thought he saw the glint of captain's bars.

He asked Fricker, "Man, I was whispering, wasn't I?"

"I don't know. I heard you pretty good. What did his mother do, anyway?"

J.W. took a breath to go on in a whisper, but First Sergeant Cowsen commanded the platoon, a fire team at a time, to their feet and into a crude wooden shelter deep in the Benning woods.

Once inside, Sergeant Poliak rushed in and smirked, "Rangers, prepare for chemical warfare." Without affording them time to look up from their stupor, to say nothing of giving them a chance to pull from their rucksacks gasmasks that should have been hanging on their web belts, he shouted, "GAS!" and pulled the pins on half-a-dozen teargas grenades, flinging them at the feet of the scuttling Rangers.

A crystalline vapor spewed belligerently from the little bombs, the musty air of the tiny cabin quickly saturated with caustic CS gas. The cadres were masked in less than four seconds.

The chemical burned the chafed skin of J.W.'s neck, where he had been commanded to dry-shave as punishment for having yelled back at the awful voices. Above the hiss of the grenades, Sergeant Poliak, mask in place and breathing comfortably, launched into a lecture on the nature of CS gas. He spoke reverently. "Rangers, Professors Corson and Stoughton were chemists who discovered this non-lethal gas forty years ago. It's called CS after the first letters in their names. Lot of people don't know that, Rangers."

With equipment scattered over the dirt floor, the Rangers pulled on their masks. But the hastily donned protectors leaked, the rubber as dried out and hard as the tires at Victory Lake. With the screaming of the cadres, J.W.'s cluster jumped into ranks coughing, eyes dripping.

Several instructors, Captain Vock among them, all wearing well-fitting, modern masks, positioned themselves in front of individual Rangers and shouted, "Mask off," then kept that cadet at attention until a sufficient number of questions had been answered to insure the Ranger had sucked in liters of the acrid toxin.

As each troop began to list, primed to collapse, the cadre questioning him snickered, "Outta here, Ranger."

Vock chose J.W. and hovered far longer than had the other cadres. By the time Vock asked his first question, J.W.'s eyes had swollen to a point that he could not focus, though he was able to appreciate Vock's furious stare. The full inquisition did not begin until J.W. had nearly exhausted the last deep breath he'd saved before peeling off his mask.

"Did you graduate from high school, Ranger?"

"Yes, sir. And I graduated from college, too."

"Yeah, where?

"Sterling College," J.W. panted, dissipating the last of his precious savings on the absurd question. Trembling, out of air, he gave in and sucked a tiny wisp of poisoned gas. It burned so fiercely, his throat gagged shut, though he was not concerned, for it was at that point the other Rangers had been allowed to escape. He leaned weakly to the side, mimicking his compatriots, feigning total disintegration. He turned to face the door, readying himself for the escape.

"I never heard of that place, Ranger."

"No, sir? Great school, sir," J.W. sputtered, his eyeballs burning as if they had been soaked in Fels Naphtha soap.

"You one of those fraternity boys? Wasting your time drinking and carousing like a dog in heat?"

"Yes, sir. I mean no, sir."

"What the hell are you, Ranger?" he demanded.

"I'm a proud Ranger, sir. R-A-N-G-E-R, sir." With the last letter, J.W. dropped to his knees, searching the dirt floor for a pocket of uncontaminated air, but it was worse the lower he went, Professors Corson and Stoughton having designed their brainchild to sink into crevices and pool in bunkers, not to rise and evaporate.

Vock carped over and over, "On your feet, dud," but the gas had crippled J.W., and he crawled in circles, the searing flood of tears pouring from his eyes forming little puddles of mud. Vock screamed furiously, "Get your sorry ass out of my sight," and J.W. lunged for where he had last remembered the door.

"It's gone. Bastards moved it," he cursed, crawling on his knees.

Blind and suffocating, he bashed his arms and head furiously against the planks until a section of rotted wood splintered. He crawled through the hole, exiting into the midnight air to dive head first into a freezing stream. Despite the bath, for hours afterwards, the staff sergeant commanded him to stop coughing, and several days passed before J.W.'s eyes lost their deep crimson cast.

"The red badge of courage," Gillette laughed.

There was still an occasional whiff of tear gas puffing from the folds of his filthy rucksack when Class 68-B dragged into Harmony Church at the end of the first three weeks. They stood in a loose, unchaperoned formation. The student leader looked about for cadres, for guidance in the final hours of the First Phase, but Harmony Church was apparently devoid of all cadres.

J.W. became bored and left ranks for the barracks latrine to spend the serendipitous interlude on the throne, a peaceful yet exciting prospect, after weeks of squatting over pricker bushes. He had not become a victim of commando crotch, the pustular bullae of poison ivy, oak, and sumac that had claimed several of the recently departed,

J.W. entered the latrine with a jaunty step. The same five, backless, shiny white, porcelain toilets stood exposed like fresh, perfectly spaced, military gravestones at Arlington Cemetery.

A body jumped from the middle crapper and yelped in J.W.'s face, "What the fuck are you doing in here, Ranger? Get your sorry ass out into formation."

"Yes, Sergeant Hartack."

Hartack pulled up his fatigue pants and bristled out into the main barracks, buttoning his fly and screaming at the troops who had followed J.W. into the billets. Vock appeared from inside the command shack and made his way down the ranks, curiously quiet as he turned the corner to inspect Second Platoon, Second Squad. J.W. had not seen him since the gas chamber, and there was gossip that Vock had taken emergency leave to have his mother committed to a lunatic asylum in San Francisco.

That didn't surprise J.W., who had learned the truth about Vock's mother from a Ranger who had graduated from Virginia Military Institute. So right before dawn that morning, J.W. had filled in the gaps for Fricker. Gillette was listening as well. "The legend is that the CO was a tactical officer at VMI. They called him Adolph Vock 'cause he was like a hanging judge in court marshals."

Gillette interrupted. "That's courts marshal. The pleural is c-o-u-r-t-*s* marshal."

"Thank you, Commander in Chief. May I continue?"

"Proceed."

"Anyway, his mother traveled by bus to VMI and planted herself outside the window of his office. He had locked it or something. When he didn't come out, she took off her blouse, but not her bra."

Fricker piped up, "That's good. She's too old for that sort of thing. No one wants to see that."

Gillette interjected, "I don't know, Ranger. Older ladies can bring experience to the table, or the bed, you might say. I mean, I don't know firsthand, but it's what I've heard. What I'm saying is, you shouldn't dismiss anything on the basis of unproven rumor."

"Jesus, Gillette, that's off the subject. Anyway, Vock hates his mother, hates getting married, and hates married Ranger cadets. At least, that's what I heard."

Fricker was nearly drooling. "What about the sex thing, what you were saying..."

"I don't know, man. That could just be a rumor.

In formation, Vock stopped in front of Bearchild and glanced fleetingly at his uniform, checking it up and down. J.W. stiffened, conscious that Vock had stared for a second too long at the man's crotch. The commanding officer said nothing; instead, he shifted his gaze slowly up into Bearchild's black, sparkling eyes. Then he moved on without a word. J.W. could see him out of the corner of his eye, heading slowly toward him, marking time, toying with his prey. He tried to cover his wedding ring, but Vock looked only into his face.

The captain lowered his voice and whispered, "Ranger, when this formation is dismissed, I would like a private word with you. Report to the tac shack."

"Yes, sir!" J.W. answered, relieved at the gentleness of the request, sensing the meeting was to make up for past misunderstandings.

When J.W. sprinted to the operations shed, the others ran to the barracks to shower and leave for town. The Benning Phase had come to an end, and though the mountains were next, a six-hour pass was first, and J.W. rushed to see Vock and get it over with. He knocked on the black-stained door and was invited in without bloodletting.

Vock sat at the meager desk and spoke calmly, "Sit, Ranger." Without expression, Vock stood and walked stiffly around the flimsy, olive green, folding table toward J.W. His pasty skin was marred with a dozen shaving nicks. Though Vock's eyes were as cold blue as a malamute's, J.W. did see in them the sparkle of intelligence that had carried Vock through his years at VMI. J.W. looked away from the expressionless face but sensed Vock's lips relaxing as if to form a smile. He turned back to face his commanding officer just as Vock's right hand started to rise. J.W. was about to thrust his own hand out to shake Vock's, but before J.W. could lift his arm, Vock slapped him with a half-fist and grabbed what stubble of hair had reappeared. Vock pulled J.W.'s head back and uttered with disgust, "You are the sorriest sack of shit I've ever had in this school. You've been a fuck-up your whole life. We know all about you. I will personally ensure that you do not graduate. You will never, ever, wear a Ranger Tab. Get out of my sight."

J.W.'s mind fogged from the semi-punch, and his eyes moistened, but with the pluck of having gotten through the fall from the Ranger Tab at Victory Lake, the Slide for Life, and the gas chamber, all sense deserted him. He lunged toward Vock. "I'll kill you, mother fucker, with my bare fuckin' hands, and then I'll bite out your mother fuckin' blue eyes!"

Vock jumped back and laughed. "Try it, dud. If I don't snap your neck or blind you for life, you'll spend the next fifty years in a military prison gettin' butt fucked. Now, get the fuck outta here. GET OUT!"

As the warning soaked in, J.W.'s head dropped and he managed to wobble from the shack toward the barracks, trying to hide his face from a few of the men who were already in civvies, trotting toward the road that lead to main post. With each step, J.W. accepted more deeply what he had been promised as a child, that he would someday reckon for his crimes. He would not avoid retribution for the embarrassment and shame he had brought upon his family, for all the mistakes, the poor behavior, the trouble at school, in sports, and even in the Boy Scouts. He remembered clearly his mother swearing his perverse behavior would elicit payment in the end, but it was not until that moment that he finally believed her manifesto of fire and brimstone, that he was, in his core, loathsome.

Though punished for shooting his brother in the ear with a BB gun, setting fire to the neighbor's lawn with a cherry bomb in a can of gasoline, producing awful grades throughout his schooling, and then fomenting a riot at Sterling College the night he trumpeted a ram's horn from his dormitory window, over the years he had convinced himself that her words were false, for he had always been able to talk himself out of trouble. He had heard of real sins, far worse than he had ever dreamed: pals with records for street sign theft, pizza theft, condom theft, and sneaking into movies—real criminals who were living quite comfortably at home, with good jobs and warm futures. He could not fathom why he had been selected to pay.

But as he continued his stagger, he began to respect his mother's wisdom and decided that perhaps Vock had been designated by fate to even the score. When his jaw began to throb, he closed his eyes and applied gentle pressure to the mandible. It helped until he stumbled into a pothole, the fall triggering a blast of lancinating pain that coursed through his head so brutally, it overshadowed the agonizing electricity radiating into his right leg.

By the time he dragged himself into the barracks, the last of his compatriots were pulling up zippers and bouncing jauntily past him for main camp. Branch was in black pants, black Banlon, and pink shoes.

Only Bearchild remained home, in his drawers, supine on his bunk, staring up at the World War I, raw-pine rafters. The wood had reddened over the years, pockets of petrified amber sap collected in droplets along the lower edge. J.W. worried the ancient roof was leaking, then wondered how many GIs had rested on those bunks watching those same beads of orange, waiting for one to fall. How many were still alive? How many had survived both

World Wars and Korea? How many Rangers had already died in Viet Nam? He tried not to think about how many were going to.

"Hey, Bearchild, how come you're not going to town?"

He grunted, "I've got better things to do with my time than get drunk." He went back to *The Two Vietnams,* while J.W. dropped onto his own bunk and slept.

It could not have been ten minutes before J.W. was startled awake by the alcoholic cursing of his cohorts. According to the grumbling, they had been gouged for steak dinners in Columbus then repulsed by a panoply of snotty local women, from waitresses to innocent customers, all of whom had ignored their propositions and shaved heads.

Branch walked by J.W.'s rack and stared at his face. "What the hell happened to you?"

"Nuthin'," J.W. grunted without moving his jaw.

Branch's once-gleaming patent leather shoes were now clay-red with dust swirls, his black Banlon a deep henna. J.W. rolled onto his side, facing away from the returning troops, wrapped again in the distant depression he had known for so many years. He was sure everyone knew of his humiliation and had spent the evening laughing at him.

The troops, however, ignored J.W. Weathersby and everything else in the world, other than settling on their racks without pulling off their civvies or brushing their teeth.

Lights dimmed, and J.W. fell into a nightmare of the entire company stoning him for having brought shame to Class 68-B. It went on for what he was sure was a full night, though seconds after he'd fallen off, he was jolted awake by the door banging open and lights flashing.

Cowsen stood in the middle of the floor. In a curiously quiet, humane tone, he drawled "Out of the sack, gentlemen. Time to pack duffels for the trip to Dahlonega."

Fricker asked, "Please, First Sergeant, could we wait till morning?"

Cowsen grunted, "Ranger, by sunrise you'll be there."

⸺✠⸺

Cowsen directed the Rangers of Second Platoon to pack all their cold weather gear for the Mountain Phase. That struck Morelotto as curious. He grumbled indelicately after the first sergeant had left to pester Third Platoon, "Man, that's the same shit I been lugging around for the past three weeks, and I still froze my ass off. He's full of crap."

Cowsen, however, flew back into Second Platoon's barracks and stood by the door, searching. Though Fricker tried to signal Morelotto, the still-pudgy man's head was buried in his foot locker, his arms flinging red-mud stained, olive drab clothing over his shoulder, burping profanities with increasing volume.

Fricker finally yelled, "Hey, Sergeant Cowsen, are the mountains colder than Harmony Church?"

Cowsen took a step into the bay. "Rangers, it's only gonna get worse, and you ain't gonna have me to protect your sorry behinds. Look here," he added, stepping up to Morelotto, "Dahlonega's hundreds of miles north. There's lots a snow in them mountains. They's thousands a feet higher than where y'all's standing right now. Y'all been to college. Y'all tell me if it's gonna be colder."

In minutes, Second Platoon was drummed into formation, dragging stuffed duffel bags. Morelotto was still carping, "Why do I have to hump this useless crap?"

At the mess hall, the staff sergeant urged happily, "Morelotto, why don't you shorten the trip to the embarkation point by sprinting. You carry Fricker's stuff, too. He's got a bad shoulder."

Waddling on the run under his own fifty-pound duffel, J.W.'s helmet popped off and bounced forward, tripping Morelotto, who was so incensed, he dropped both bags, grabbed the helmet, and

146

flung it by the chinstrap at a parked jeep. The rivet attaching the canvas strap to the metal shell pulled free. J.W. retrieved the helmet angrily and pushed the rivet back into place with his teeth, chipping an incisor.

At the truck point, they stood shivering, eighty-some men, the victorious final half, hopping up and down against the bitter cold. Fricker suggested that perhaps the cadres would find some excuse to make them walk to Dahlonega. A thousand arguments were proffered dismissing the notion of such a march. The chatter flew about like bats in the night, the movement of jaws apparently generating soothing body warmth, for it went on and on. But as they waited for the cattle trucks, several of the Rangers began trembling at the thought of a death march.

An hour passed, and without the appearance of the cattle trucks, Fricker warned through chattering teeth, "Accept it. Anything's possible. They're gonna make us do it on foot."

When he spotted a cadre his hand began to rise, but Branch and Bearchild grabbed Fricker's arms and held them by his side. J.W. agreed, silently, that it was better to contemplate the agony of a walk to Dahlonega than actually take the first step. Though he had learned in Psych 101 that the human mind was incapable of distinguishing between reality and a belief deeply held, after three weeks of Ranger School, J.W. knew that theory was badly flawed. He would have waited there for one month, in the bitter cold, thinking about walking, fretting about walking, rather than begin the tramp.

Another half-an-hour passed, and four empty trucks bumped and bucked up to what was left of Class 68-B. Morelotto grunted, "Those are the same goddamn trucks that 'a been parked behind the mess hall since yesterday afternoon."

Fricker, when he was sure Branch and Bearchild were not looking, raised his hand and asked, "First Sergeant, why couldn't we have waited in the trucks? At least we could have been warm for a while."

After the fifty-first push-up, the defrosted troops gratefully flung gear aboard and crawled onto the vehicles, pawing for seats on the wooden sideboards. Bearchild, last over the tailgate, saw there were insufficient places, and settled onto the truck bed, arranging several of the duffel bags into a queen-sized mattress. He took *Idylls of the King* from his pocket and, by the beam of his flashlight, escaped into the pages. The rest of the Rangers drew closer to each other to capture what heat they could.

The ride into the Smoky Mountains to the Ranger Mountain Training Camp in Dahlonega, Georgia, degenerated immediately into a drill of seeking to remain upright, Rangers burning more calories in the trucks than at the pull-up bar. The suspension of the army cattle trucks had been engineered, Gillette suggested, by the same designers who forged the M-60 A-1 E-1 tank, a design that safeguarded one-hundred-five millimeter cannons, not the lives of emaciated troops.

Every two seconds they struck a chuckhole, and Branch's compact body launched a foot or so above the unpadded bench then snapped back onto the wooden plank as gravity reasserted itself. J.W., still forty pounds heavier than Branch, levitated only a few inches, but his jaw rattled sorely with each landing.

After a few miles, Bearchild propped his dying flashlight against the tailgate and read happily until they struck a Manhattan-class pothole. Though his torch flipped out the back and J.W. yelled for the driver to pull over, Bearchild had already rolled on his side, covered himself with his cold rubber poncho, and fallen asleep. The truck rattled on into the night.

When J.W. awoke, the air had become even colder and harsher than at Harmony Church. He watched the stars from the open-bed truck, laughing that the canvas tops were stowed only when it was terribly cold, raining, or snowing, and tonight, the tarps were somewhere, but not serving the passengers.

After several hours, their truck began a gentle, northerly climb, and as the driver downshifted, the Rangers crowded even closer for warmth. With each lower gear, the temperature slid another few degrees south. In the bare light from the truck behind, the troops on the opposite bench appeared huddled masses, and J.W. whispered to Gillette, "They look like my grandmother and her family when they sailed from Poland to Ellis Island in 1911."

"The only difference," Gillette snarled, "is that your grandma was cleaner, better fed, and less diseased."

The grade soon became so steep the trucks were forced to use low-low to keep the load moving up the mountains toward Dahlonega. In that gear, every twitch of the driver's foot jammed J.W. against the wooden side rails. Part of Bearchild's mattress flew over the tailgate, but he caught the duffel bag and dragged it back aboard. J.W. tried to light a smoke, but his hands were so numb, he could not make his fingers pinch the match.

An hour later, J.W. was awakened by a lurch to the left and then a thud. The deuce-and-a-half jolted to a halt. J.W. thought he heard the engine growl, though it snarled as well, and he realized it was animal noise, not the demise of the diesel motor. There followed a vicious gurgling, a weaker growl, and then the silence of their first mountain dawn.

J.W. peeked over the side rails to see a black, amorphous mass under the left front wheel of the cattle truck. He assumed it to be a grand, moss-covered stone that had rolled off the craggy mountains, but a tortured head thrashed out of the glob, twitched a time or two, then fell back to the macadam.

The truck rolled a few feet in reverse, allowing the massive lump to unfold. Fricker leaned over the rails and gasped. "It's a bear!"

The thud reminded J.W. of the time his mother had run over the neighbor's collie, though this thump was far more sonorous, as Laddie Boy hadn't weighed four hundred pounds.

The staff sergeant pulled up in his heated jeep. "You, you, you, you, and you, dismount and pull that thing off the road before the goddamn Fish and Game man happens by."

As J.W. stood to dismount, he played the accident over in his mind. "Hey, driver, tell me how you managed to hit that thing. It was in the left lane, and we were on the right. How could that have happened, Private?"

The man glared at J.W. "Ranger, mind your own fuckin' business."

J.W. and cohorts approached the carcass. Though the creature was motionless, he remembered from third grade that wounded animals could not be trusted, so he just pretended to help. He watched bugs flit in and out of the bear's moth-eaten coat and observed the deceased's yellow teeth, thick with tartar, just like his. The odor was hideous, and J.W. gagged. He turned away and faced the mountains, beholding for the first time the magnificence of the Smokies.

A light mist hung above the craggy hills, summits capped with a delicate snow so white, the peaks glowed in the first rays of dawn as if it were noon. The highest, Brasstown Bald, was distantly monumental—Class 68-B had been transported in just hours to the most magnificent point on Earth. The men were silent in awe, even Branch. J.W. fixed his gaze on the lesser hills, nearly as breathtaking. He looked forward to touching their splendor.

# CHAPTER 2
# MOUNTAIN GUERILLA

Ranger Mountain Training Facility
Dahlonega, Georgia
Deep in the Smoky Mountains

Once parked at the Ranger mountain camp twelve miles north and east of the Dahlonega, they were marched to a line of huts, primitive, ten-man shelters, sturdily built of local, rough-hewn lumber that had weathered decades in the rain forest. The floors were thick pine worn smooth by Ranger forefathers, black dirt from combat boots ground into the wide gaps between the boards. It had hardened over the years, chinking out the piercing cold.

J.W. inspected the facilities of the squad bay. The bedsprings were coated with a flaking layer of dull rust. The constant moisture had also fostered a cottony patina of mildew on the paper-thin mattresses, but the scent of his single-celled companions blended with the surroundings, and J.W. allowed it was really quite cozy in its own way. Though drizzle fell outside, and they would soon be wet on top of cold, there was a permanence to the camp, a rugged-ness that comforted him.

He went to the back door to examine the latrine. It was not as inviting as the sleeping quarters. Judging by the odor drifting from the cobweb-covered holes in the rotting plywood, they had also been there for the millennia. A brimming No. 10 can of pungent, off-white lime sat officiously by each hole, worrying J.W. that his predecessors had been worked so hard, they'd lacked the energy to scoop into the cesspools a few ounces of the lime that would have sweetened Dahlonega's ambiance.

He returned to his bunk, quietly enjoying the few moments of peace, listening to the rain beat on the shake roof. He savored the dry softness of the musty bed after the hours upright on a rock-hard bench and pulled from his duffel bag a book Bearchild had loaned him, *The Agony and the Ecstasy*. He couldn't wait to cover himself with his poncho and read, but after just a few sentences, he was asleep.

His dreams ended in minutes, exterminated by the boot steps of a mountain phase cadre, another staff sergeant, who marched into the hut and bellowed, "Fall in." J.W. stood in front of his rack and looked out through despairing eyes into a lean, stoic, perfect, Teutonic face. After the blond hair and blue eyes, J.W. noticed the name tag, GEHRING, embroidered in thick black over the sergeant's breast pocket. Weathersby, still in his near coma from three sleepless, foodless weeks, fantasized it was 1943, and that he was a POW in Germany, but when he stared harder at the name, he realized the spelling was different. And anyway, when Weathersby looked down at his own red-stained, shredded fatigues, he wondered what Hitler's second in command could want with the likes of J.W.

The man drew a deep breath and boomed, "Pack for the first real tactical problem of Ranger School," then paced the gritty floor for a few seconds, evaluating the new crop of subjects with a discerning eye. Sergeant Gehring abruptly spun about and marched through the door. The Rangers were on their beds before the man's boots hit the ground, then back up on their feet when he stuck his head through the door and delivered the most significant part of his message, "And no goddamn candy or any of that kind of crap in the Mountain Phase. Got it?"

He left for the next hut and, seconds later, J.W. heard from that squad's billets, "And no goddamn candy or any of that kind of crap in the Mountain Phase. Got it?"

In formation, Class 68-B dropped their fatigue pants and touched their toes, languishing in that position while the first student platoon leader was called aside and instructed in the art and science of the butt check for contraband. He was left with the message, "Blessed is the Ranger who discovers a candy bar hidden deep within his comrade. "There's extra points for each Almond Joy you extract, Ranger. And you even get to eat half of anything you find."

There was, in fact, no way to sneak candy past the despots of Dahlonega. Claiming to be past masters in the detection and punishment of smugglers, they bragged, "Rangers, we've seen and crushed every trick you can think up to beat the system. Don't even try. It's gonna cost you if you do."

And the cadres discovered hiding places J.W. could never have imagined. Fricker, who had come up with the ploy of secreting a Hershey Bar in the hatband of his fatigue cap, was discovered when a cadre came up to him from behind and smacked him upside the head. The already-melted chocolate dripped out and mixed with the grease in his hat and on the collar of his fatigue shirt. The one survivor of the cadres' scrutiny was Bearchild, who held a Mars Bar in each hand during the purge.

Sergeant Poliak, chairman of the machine gun pirates and one of the few carryovers from Harmony Church, belched, then exploded in Fricker's face. "Ranger, couldn't you figure out you were wearing a billboard saying you were cheating? Rangers don't cheat. So, first you eat the wrapper, all of it, and if you're still alive after the tinfoil, and most Rangers don't live through it, I want you to suck on the hat brim and your fatigue blouse until the last smudge of brown is gone. Ranger, when you're done, the only thing left on that hat better be the grease stain that was issued with it."

Poliak then sank into a curious silence, issuing no threats of push-ups or even curtailed meals as Fricker made smooching sounds on his uniform, slowly slurping the body-oil-enriched chocolate. Poliak's only words as he strode to the front of the formation were, "Least you had the balls to try, Ranger."

The company marched along worn, meandering paths, past a frozen brook, coming finally to the training Quonset huts of the mountaineering school. Waiting patiently inside on the floor lay dress right dress heaps of sinister hardware, chunks of black and silver metal, and robin's egg blue, yellow, and burgundy coils of nylon climbing rope. Bearchild eyed the collection and nodded knowingly. So did Gillette.

Sergeant Gehring took the podium. "Rangers, welcome to Dahlonega. When you depart, you will have the basic skills of mountaineering. That's for sure. But more important than the hundreds of miles in front of you, Rangers, is that you will have learned what it means to never stop, no matter the mountain in front of your eyes. This isn't the Benning Phase anymore, you know. This isn't good food and a warm bed. This is the training you were sent to Ranger School to master. This is the real thing. This is where you will learn to drive on."

With those two simple, single-syllable words, J.W.'s heart popped by reflex out of his chest and into his mouth. Then his bowels spasmed, and though there was nothing inside of him to lose, he searched with his eyes desperately for a latrine.

"Gentlemen, anything is possible if you just do it. You already know that concept, I hope. It's the 'just doing it' that we're going to teach you here for real. We tolerate no excuses. Every time you hear the words 'drive on', you will do just that. DO IT! You will drive on and on and on."

J.W. stood and looked left and right for the john, the chances of averting a terrible accident slimming with each "on" Gehring howled.

The sergeant clucked at J.W., "Sit down, Ranger. You got ants in your pants or somethin'? Gentlemen," he continued, "you see before you the tools of your trade." He held up each item and called out in a crisp, military voice, "Swiss seat—D-ring—belt—rope—piton—crampon," though all J.W. could think about was finding a crouton.

Sergeant Gehring pointed to J.W. "Hey, Gunga Din, you like bein' on your feet so much, get your butt over here." He stared at J.W.'s nametag. "Hey, you're the one who said he could climb for-ever, right?"

"No, that was 'swim,'" Fricker volunteered from within the olive drab mass of dozing troops.

"Who cares, Ranger?" Gehring handed J.W. a fifty-foot, thirty-pound, nylon climbing rope. "Put this around your neck and stand in the corner, Ranger. You wait there until instructed to move. And you, Ranger," he called to Fricker, "you hump the bag of rugs for your trouble."

J.W. walked to the appointed position, commenting through gritted teeth under his breath, "Yellow rope. Clashes with my red-mud-stained fatigues." Branch nodded in solemn agreement, and J.W. felt, for the first time, the support of a friend, one who might even take a turn carrying the line.

Then Branch quickly added, "Yeah, but it matches your personality."

The rope's sharp nylon bristles dug through the raw skin of J.W.'s neck until blood seeped into the collar of his fatigue jacket. He did not want to sacrifice the other side of his neck and risk the possibility of the rope touching his swollen jaw, but when the blood dried and stuck to the material, he had no choice. The rope did chafe his cheek, which soon swelled. Gehring looked over and queried, "You got the measles or somethin', Ranger?"

Gillette mused, "I think you mean the mumps, Sergeant. You see, mumps is an infection of the parotid glands, while measles..."

Despite Staff Sergeant Gehring's uncertainty about the symptoms of various pediatric infectious diseases, there was no misunderstanding as he ordered the first fifty-and-one of the half-million push-ups to be executed in the mountains.

Arms heavy with the blood of exercise, the company, loaded as if Sherpa bearers, marched to a set of rotting bleachers. There they were tutored in the procedures vital for climbing and surviving the challenging and dangerous mountains of the northern Georgia rain forest. That took six minutes, after which they were each handed an eight-foot length of rope from the crate Morelotto had lugged. "Roll it exactly four-and-a-half times, place it in the left front pocket of your fatigue pants, and fall in for the first exercise of Ranger School."

A short march farther into the mountains brought the company to a fifty-foot wide, thirty-foot tall wall of black-painted wooden planks. A mammoth orange Ranger tab was stenciled across the shiny black of the vertical lumber. Stretching across the top of the structure sat a narrow wooden platform, a terrace, upon which a squad of instructors glared down, hawks eying prey.

After brief remarks regarding rappelling technique, six Rangers at a time were sent to scrape up the hill behind the wall, then leap across the three-foot gap onto the thirty-foot high balcony. J.W. was number one.

Gehring snorted to J.W., "Ranger, take the eight-foot piece of rope we issued you and tie a Swiss seat." Without waiting for a retort, he went on. "Don't know how, huh? Weren't you paying attention?"

Though J.W. had tried hard to remain conscious during the mini-lecture, in the five seconds he'd let his eyelids droop, he missed seventy-five percent of the twists and turns.

He fumbled through the first part of the process, and the sergeant snapped the rope away from him. "Watch me closely. I'm only gonna show you one time."

The sergeant's deft fingers wove and looped the rope through his own groin several times, fashioning a mountain climbing sling that circled around waist and butt. "This is the famous Swiss seat everybody's heard about."

Gehring handed the slip of rope back to J.W. "You tie it!"

When J.W. failed to get it right on the third try, Gehring snatched the rope away and tied it for him, cautious not to allow his fingers anywhere near J.W.'s pelvis. An air of revulsion blanketed him, one he made quite evident to his audience, as his hands neared the front and rear portions of J.W.'s crotch. He snapped two D-rings onto the climbing harness at J.W.'s bellybutton, placed a loop of nylon rope through the rings, instructed J.W. to stand at the edge of the platform, backwards, and prepare for his first rappel of the Mountain Phase.

"I don't know how to rappel, Sergeant," J.W. moaned.

"We just gave you a full class! What is wrong with you? One more time. Turn your butt toward the valley, feet halfway over the ledge, and lower yourself backwards until suspended parallel to the ground by the rope in your left hand. Use your right hand, the lower one, on the rope behind your back to control how much line you let out. Swing the right hand forward when it's time to descend, and snap it behind your back smartly when it comes time to stop. That's your brake hand, Ranger. Don't forget it.

"Okay, now walk down a few steps backwards along the wall. Push off a little, that's it, let a little more line out. And try to move faster than one inch per hour. Dinner'll be over if you don't get a move on.

"And when you get to the ground, yell, 'off rappel!' Now, is that so hard? And open your eyes, Ranger. You're supposed to be looking for the enemy at the same time."

J.W. lowered himself backwards, a millimeter at a time, praying as the rope became increasingly taut that the thin nylon cord would support him.

Gehring nodded. "You're doing good. Keep going. You're getting close to being parallel to the ground, Ranger. Let out a little more rope with your right hand, just walk down the wall, and when you're six feet from the bottom, stop and rotate into the feet-down position. It's that simple."

J.W., though, asked himself how he was supposed to know where he was with his eyes so tightly locked shut. Gehring huffed then massaged his student with a few volleys of screaming threats, but J.W. remained frozen, a red-stained, OD dowel hanging nearly head down from a pegboard twenty-nine feet off the ground.

Gehring bent forward and plucked the rope hard—the vibration was loud and low like a string on an upright bass fiddle. It set J.W. shuddering, and his boots lost their precarious traction on the muddy boards. His feet shot upward, his head pointing straight down, now twenty–five feet off the deck.

That opened his eyes. He hung there until Gehring advised him to play out bits of rope, which he did miserly, but wriggle as he might, nothing worked. Gehring sighed loudly and leaned over the edge with a Bowie knife in hand. That brought J.W.'s brake hand forward, and he was soon sailing toward the earth, but still head first.

He gathered speed, the air whistling by his ears as in airborne school, and at the Slide for Life, though this wasn't a two-thousand-foot aircraft drop with time to prepare for touchdown. And he wasn't upside down during those exercises. He squandered altitude wiggling and gyrating, and when he extended his neck to assess his altitude, the ground was moving up toward him at warp speed. He closed his fists in a death grip on the line, though both hands were in front of him. He kept falling. Friction from the rope burned through his gloves, and steam puffed from the soaking, wool glove liners.

"Use your brake hand, your brake hand, Ranger!"

More steam.

"Throw your right hand behind your back!"

When he puzzled out which hand was which, and by chance chose the correct one to throw behind his back, there was a bang that could have halted an armored personnel carrier. His torso came to a stop so abruptly, his legs swung into the foot-down position inches from the ground. He loosened his brake hand and settled to the frozen dirt. He screamed, "Off rappel," unhooked the D-rings, ran to the top rung of the bleachers, crawled into a little ball, and smoked three cigarettes.

After rappelling came mountain climbing. They marched to another set of bleachers, one at the base of a nearly vertical slab of mountain. While J.W. remained awake for the exposition on how to scale granite walls using just fingers and toes, he couldn't believe that was how real climbers did it. Gehring watched his eyes and sent him to be first in line, again. The sergeant commanded, "Ranger Weathersby, use your hands and the tips of your boots like I just told you. And stop at the forty-foot level."

Fricker's hand shot up. "Sergeant, excuse me. I don't understand. I didn't know you had to use your hands to climb a mountain. I always thought the next higher guy pulled you up by a rope or something."

Branch bent forward into the push-up position as Gehring sighed, "You must be Ranger Fricker? And, what the hell are you doing down there, Ranger?" he moaned at Branch. "Lookin' for money? Well, while you're down there, knock out ten-and-one. And it ain't a gift. We just don't want to tire out your arms."

"Yes, Sergeant," Fricker answered proudly.

"You wanna do 'em, too?"

"No Sergeant. I was just saying that I'm Ranger Fricker."

"Okay, everyone on your feet—you, too, money man." Gehring clacked, "Start climbing. I want all of you to stop on that wide granite ledge up there."

The outcropping about which Gehring crowed was twenty-five feet above them and three feet across. J.W. remarked under his breath that it was wider than the shelf on which his first Ranger buddy had maimed himself. It was sufficient, though, for a platoon to squat in a huddled formation, which they did ten minutes later.

"Rangers, you are about to learn the Golden Rule of Mountaineering: you are instructed to maintain three points of contact at all times. I don't know if they taught you that at Benning, but I'm teaching it to you now."

J.W. turned to Branch and whispered, "What was his name, Smith or Jones? I wonder if he made it?" Branch shrugged.

Gehring roared that the class was to climb higher, to the next level, which was atop a nearly sheer granite face. He reminded them about the value of maintaining three points of contact. Though J.W. sought to pay rapt attention, his concentration was broken when hair-raising screams blasted from the summit of the final rock wall. Three cadres stood on the precipice a hundred feet above them, howling at the Rangers, jumping about like incensed mountain gorillas. The instant Sergeant Gehring pumped his hand in the air, the triad launched into a headlong, face first dive.

Though attached to mountaineering ropes, those cables remained slack, for the flying Dutchmen sprinted down the face of the cliff, bodies parallel to the ground, moving faster than gravity could pull them.

For the entire plunge they whooped, "Drive On! Drive On! Drive On! Drive On! Drive On!" until, within a few meters of crashing into the Rangers, the climbers violently threw their brake hands behind them and touched down softly on the ledge next to the hopeful commandos. J.W. did not see them stop. He was busy searching for a latrine.

Gehring smiled. "Rangers, if there had been a dozen fresh eggs on this ledge, those Green Beret's wouldn't have crushed them. And because of that, there is no way to crack eggs here in the Mountain Phase. You won't be getting good, fresh food like you did in Harmony Church. And you won't need it. You are obviously carrying around enough fat to survive in these mountains for several months if you get lost or miss a ledge."

Fricker raised a tremulous hand. "What happens if you *do* miss a ledge?"

"Ranger," Gehring nodded, "that is a reasonable question. If you miss a ledge, your rappel will terminate in the woods, fifteen hundred feet below. But that will not happen because we make sure no one ever falls in Ranger School."

Several of the troops turned gingerly, to peep into the valley, though J.W. remained in a squat, his back to the gorge, stare fixed on the rock wall two inches in front of his eyes.

Fricker whispered, "What the hell is this fascination these guys have with altitude? They must be frustrated pilots or something."

J.W. shook his head and hunkered more tenaciously, forehead leaning against the rock wall, his eyes tightly shut, thankful for the feel of the cold granite against his brow.

Gehring turned toward J.W. and shook his head. "Ranger, step backward away from the wall." The command sent J.W. into a deeper crouch, and despite the freezing rivulets of spring water that soaked into his fatigue jacket, J.W. leaned with more pressure against the rust-tinted, sandy layers between the granite. Pushing Rangers out of his way, the sergeant strode along the narrow outcropping until he stood directly over J.W.

"Ranger, gimme fifty, NOW, and with your head over the side of this mountain!"

J.W. considered the usual option of standing and screaming, "I quit this shit," but realized the sergeant had said "fifty," and not "fifty-and-one," and that made J.W. happy. He considered standing and walking arrogantly to the edge like a real man but instead crawled on all fours, willing to endure the laughter and ridicule of his peers, a small price for the relative security of proximity to Mother Earth. At the precipice, shame overwhelmed him, and he took a deep breath, shooting a glance into the valley. It was indeed beautiful, awesomely so, he nodded to himself grudgingly, and he sought to maintain that admiration of the terrestrial sphere in his heart.

If he fell, he laughed to himself, the last thing he would ever hear in this life would be a dozen cadres angrily ordering him to stop falling and finish his punishment like a Ranger. The notion that the cadres could yell all they wished and he didn't have to obey gave him great succor, until Branch waddled over and whispered, "It was Smith."

J.W. ruminated about the maelstrom in which he would find himself if he died, and the cadres were dispatched to reclaim his body. He pictured gaggles of livid sergeants taking breaks, drinking coffee from thermoses, and cursing while busting the nearly impenetrable web of leathery-leafed rhododendron and mountain laurel. He wondered, though, if they would be so inclined to search, or perhaps simply wait until a lost Ranger in 1974, 1981, or maybe 2018 discovered his remains.

J.W. commenced the push-ups with strict attention to form, counting aloud in military cadence, holding his back perfectly straight, dropping his chest to touch the sharp granite ledge, and returning to the arms-extended position with an explosive snap of his arms. His earnestness would impress Gehring, and the sergeant would soon grow bored and turn away. As long as J.W. could be heard counting cadence, he was safe, and when Gehring took to pestering his other serfs, J.W. switched reflexively into the surrogate Ranger push-up sequence, one characterized by arms remaining straight, but head bobbing in rhythm with the count. Each dip of the head was accompanied by a more emotionally anguished groan.

"Twenty-threeee, twenty-fourrrrrrrrr." J.W. surprised and impressed himself with the authentic strain in his voice. He kept his head rotated to the right, watching the sergeant, ready to execute the genuine article if Gehring twitched a millimeter to glance back. They had already been schooled in the precept that a Ranger never gave up the element of deception.

At push-up number thirty, however, the air changed. It wasn't Gehring—he was still badgering the others to use their fingers to climb the rock wall. It was something else, something suddenly inharmonious, a disquieting static not unlike that surrounding the glowing high-tension wires back at Harmony Church.

He shivered as he considered his options. If he looked to his left, toward the hostile vibration, he would lose sight of the sergeant, a lapse a crafty Ranger could not consider. On the other hand, something was making him uncomfortable. He dipped his head, allowing only his eyes to strain left. They did not drift that far before the image of widely-bloused fatigue pants tucked into spit-shined jump boots burned painfully into the fundi of his wide eyes. The bayonet-like crease of the starched fatigues moved an inch closer to his face, and in the reflection of the mirror-like boots, J.W. recognized a truncated face that culled memories of terror and recurring nightmares—his own.

Though he hadn't needed to look further, he could not stop his eyes from rising beyond the Brassoed belt buckle to the interloper's nametag. J.W.'s next push up was executed in perfect military fashion, devoid of pain behavior, in crisp, shouted cadence, as if only the first of many to come. He sensed both fear and security all at once, then a heavy, gripping sensation in his chest, and finally a peculiar comfort. It took him back to the Bronx and the day he finally confessed having stolen a Mars Bar from Mr. Boyle's candy store when he was eight. In an instant, however, the feelings of serenity disintegrated into a terrible black cloud, and J W. was overwhelmed by a sense of impending doom. Whatever the pain of the push-ups, that was going to be less, he grasped, than what awaited him when they were over.

"You just ain't gonna change, is you, sorry Ranger?"

"No, Sergeant Cowsen. I mean yes, Sergeant Cowsen, sir."

"Don't call me sir!"

J.W. finished the thirty-fifth push-up then paused in the arm-extended position, resting for the final assault. He took the time to contemplate his state of affairs, wondering if this was finally the big one, if Cowsen had reached the end of his rope and would summarily terminate J.W.'s military career, and over a push-up.

"What you waitin' on, Ranger?" Cowsen asked in a disheartened tone. "Instead of bragging about doin' push-ups forever, looks like y'all gonna actually be doin' 'em 'till kingdom come. I just hope I live long enough to see that. Ranger Gillette, how many years do I have to live to live forever?"

Gillette began, "First Sergeant, that's a function of your blood pressure, smoking, your genetic profile, and…"

"Shut up, Ranger."

J.W. stared into the valley a quarter-of-a-mile below then laughed weakly. "First Sergeant, forever is how long I'm gonna be in this man's army. And it wasn't push-ups, it was swim."

"You gonna do that, too. Now hush your mouth and quit that threatening me that you gonna be in my army to the end of time. Just go on and finish your exercises."

J.W. asked himself if he had the strength to dip down for three more push-ups, if he had the will to go on just that far. He sneered at himself aloud. "Do three!" He did. "Nothin' to it. Do three more, asshole, and get that look of straining to death off your face. Just do your work." He did. "Not that hard, huh?"

No, he shook his head heatedly, it wasn't hard. It was nothing. But then it floated across his mind that it wasn't the next three push-ups, or the next three hundred. It wasn't the endlessness of the military push-ups, for certainly, they would come to an end—everything came to pass. But there was no comfort when he realized that everything did not include the endlessness and darkness that clouded his future, the promise that, after the push-ups, after Viet Nam, after graduate school, no matter what came next, there would always be his need to gain approval and respect, and his abject inability to wrest that regard from the world. It had been his life, and he did not know how to make that change, to fix it. He would always be the epicenter of the next fifty-and-one.

J.W. dipped for number forty-four, considering, as he paused to rest, the political science of Ranger School, a true democracy in which every man had a choice, voting on his own future fifty thousand times a day, with his arms and feet, with each push-up and each step. Here, success boiled down to thrusting his arms forward or putting one foot in front of the other.

Though the pain in his arms built, there was the vision of his old football coach, Lewis Smith. "Heart, Weathersby," he'd growl when J.W. hadn't a single wind sprint left in him. "His heart, that's where a man lives."

J.W. dipped for three more.

There was sufficient time between push-ups forty-seven and fifty for J.W. to consider the conversation he'd overheard between Cowsen and a younger instructor sergeant back at the Harmony Church phase. The junior NCO was bitching about the military: the meager pay, slow promotion, constant reassignment, weeks and months in the field, and deployments to Viet Nam every other year. The unhappy sergeant vowed to finish his enlistment, quit the Army, and never look back.

He chided Cowsen, "You're a good man, First Sergeant. Why the hell do you keep re-uppin'? You could do better than bein' a lifer."

Cowsen pondered for a moment then answered quietly. "Young sergeant, when I enlisted twenty-six years ago, I spoke three words of English: whoa, gee, and haw. That's 'cause the only thing I talk-ed to on that farm in Alabama was the family mule. My father was a cripple from an accident, and my mother died givin' birth to my baby sister.

"When I was ten, Sergeant, *I* was the sharecropper. *I* was the one raisin' the chickens and the hogs and that stubborn-ass mule. Now, I'm married. And the Army gave me an education, and my kid's goin' to the Naval Academy. A Negro sergeant's son is at Annapolis! Gonna be a naval officer, Sergeant! Wants to fly jets off a carrier." Cowsen paused, and his eyes clouded, but he gathered himself then whispered, "Don't you tell me this is work. Shooooot, this ain't nothin' but a vacation. I got dreams, Sergeant, and they is a comin' round. You do what y'all will. There ain't no better life for me than the army."

J.W. finished his fiftieth push-up, added one on his own for the Rangers of old, and was then awarded ten-more-plus-one for disobeying the order to do only fifty. As he crawled back into his squat against the frigid granite, his eyes sought Cowsen, not to greet him, not to smile broadly in his face and prove he wasn't yet dead, but to make sure he knew where Cowsen lurked.

The first sergeant had disappeared into the vapor, and Gehring, sensing that he was again the alpha male, barked, "Ranger Weathersby, I know you're not going to walk, so you might as well crawl front and center."

With J.W. on all fours, close enough to lick the sergeant's boots, Gehring hissed, "Jesus Christ," and turned to the company. "Okay, all of you's, up to the top, if you don't mind. And you, too," he spat next to the cowering mass hugging the thin shelf of granite.

At the summit, J.W. found the same tribe of Ranger instructors who had been on belay at the wooden rappelling wall. Here, how-ever, they were jamming heels into the shale, forming little caves into which they'd plant their boots. Some of them were rubbing gloved hands over ropes, massaging them, as if preparing to anchor grizzly bears. Half the Rangers were assigned to sit beside these men and absorb the art of grinding boots into rock and wrapping loops of rappelling rope around their bodies. The other half was directed to the pile of worn swatches of throw rugs Fricker had carried to the mountain. Gehring's eyes focused on the rug detail. He spent many minutes lecturing, more than he had invested in climbing and rappelling, on the military method of placing carpet remnants over the sharp granite edge of the ledge.

"Gentlemen," Gehring explained, "these rugs are here to protect the nylon rope from rubbing against the rock. If the rope breaks, the attached Ranger will land in the rhododendron thickets several miles below. We've gone over that before." He placed a mat on the edge, positioning it long ways and short ways, carefully considering the geometry until he smiled, "Still haven't got it right." So he arranged it catty wompus and stared at the students. He wagged an index finger. "Don't let me see you do it this way, gentlemen."

J.W. thought that a caring gesture, for the cadres to be so concerned about the security of the troops. His impression of Gehring softened until one of the Rangers stood at the edge to study the rug's alignment. He became so engrossed, his M-14 slipped from his hands, bounced on the granite ledge, and careened over the edge. As it was still connected to the Ranger by the shoelace sling, the man was yanked a millimeter or two toward the precipice. He countered by jumping back three paces. That drew the mildewed bootlace over a bald spot in the carpet. The rotted cotton shredded on the razor-sharp granite and the weapon tumbled out of sight into the rhodies.

Gehring's face took on a glow, though it transformed into seething rage before the final clunk of the rifle reached the Rangers' ears. The sergeant began pacing in circles, conjuring obscenities and insane plans for mass slayings over the paperwork he faced to explain the loss of United States Government property.

Second Platoon, every man save the troop who had dropped the weapon, was sent into the forest on a reclamation mission, an exercise that took until dark. They missed dinner, but it was better, J.W. agreed, than having to rappel.

With the platoon reassembled on the ledge, Morelotto squatted to pack his gear, expecting to be marched to the cabins. It had been rumored that once they were doing real climbing, they would be given full meals and four hours of sleep in the Mountain Phase, just enough to keep the Rangers alert. Troops tumbling willy-nilly from cliffs would eventually make it to the newspapers—perhaps one or two per class could be maimed or die unobtrusively; a dozen would surely be noticed.

Gehring bristled up to Morelotto. Shaking his head in disbelief, he clacked, "Are you out of your mind? Tie your Swiss seat and hook up to the rope. You're going to be the first Ranger of Second Platoon on rappel."

J.W. smiled and crept into the shadows toward the rear of the line, having forgotten just how adept the human eye was at detecting movement, particularly at night. Gehring stopped in his tracks and smiled broadly. "Hey, Rappel Forever, front and center. I changed my mind. You hook up. You're first, as usual. And you, Packing-Up-Your-Pack-Too-Soon-Man, assume the position and remain there until Climb Forever here completes his mission."

Gehring turned to J.W., who had tied his Swiss seat backwards. He shook his head and muttered, "I don't know, it might work. Let's just see. Okay, now turn with your butt to the valley, feet halfway over the ledge, just like at the rappelling wall. Now, lower yourself backwards until suspended parallel to the ground. And you better yell, 'off rappel' when you get to that ledge, or when you're flying past it, whichever comes first."

J.W. spoke to Gehring, voice quivering. "Sergeant, I still don't know how to tell when you're on the ledge."

Gehring reassured, "When your feet stop, that's when. Quit your worrying, Ranger. Your safety is our first concern. I told you, no one ever falls in Ranger School. Now get your butt in gear before I give you an assist."

J.W. lowered himself into the semi-rappel position in fits and starts, hanging backwards at a forty-five-degree angle, refusing to look down into the murkiness miles below. He remained frozen, half-way into the mission, his mind churning with calculations of terminal velocity and impact kinetic energy.

"MOVE YOUR ASS, RANGER!"

J.W. took a last look at the ledge he had backed over. By the flicker of lanterns, he focused on circles of the ragged buffer carpet that had worn through. New whiskers popped from his rope with each degree he leaned farther back. He lowered himself quickly, assuming the faster he finished the rappel, the less time for the rope to fail. As he placed his full weight on the rope, a crew cut of new filaments popped up before his eyes.

J.W. started down the sheer cliff backwards. Though he could see nothing, by reflex, he gnashed his eyes shut and refused to respond to Gehring, who spit, "Come on, Ranger, spring out. Give the rope slack. You got twenty of your buddies waitin' here in the cold on you, and one in the push-up position." Gehring sprinkled dirt over the side of the cliff.

J.W. pushed off, counted to three, then brutally threw his brake hand behind his back. He came to rest twelve feet down the rock wall—once again, upside down. His jaw spasmed with the sudden stop, and he called up to Gehring, "My jaw's broken. I gotta stop."

A grape-sized stone landed on his gut, rolled along his upside-down-chest, and wedged in the space between his upside-down-head and his upside-down-helmet.

He scrambled to right himself, hoping gravity would dislodge the nugget that was grinding into his scalp. When a plum-sized rock bounced off the sole of his boot, he sprang away from the wall, giving the impression of progress. The next rock hit his brake hand, and he brought it forward in reflex. That begot ten more feet of headway, and also brought him back perpendicular to the sheer rock face.

"You're gumming up the works, Ranger. Move down. NOW!"

J.W. shoved off, spinning fifteen feet of rope through his gloves, though it dawned on him only after he was sailing, that, unlike Ranger School, the rope was not forever. It had an end, one that would arrive long before the valley floor. J.W. also had no idea of what fifty feet on rappel felt like, or if there was a warning signal near the end of the rope, like his wife's, tissues which turned pink when the box was almost empty.

He braked to a violent stop and called up to Gehring, "Sergeant, I don't know where fifty feet is."

Gehring answered, "Think back to the cargo net on the obstacle course at Benning. Pretend you're looking up to the top of the bleachers. Now stop talking and move your ass."

With the next rock, he sprang from the face of the mountain and sailed with air whistling through his helmet. He grabbed the flying nylon with his brake hand, but as in Rappelling 101 at the wooden wall, he forgot to wrap the rope around his back. The nylon burned through the patch of cloth he'd cut with his teeth from his fatigue shirt to replace the hole in his woolen glove liners. By sheer brute fear, though, he managed to hang onto the rope and stop himself several feet short of a meager ledge.

Bobbing on the rope like a yo-yo, he slipped back into the upside-down configuration just as Sergeant Gehring shined a flashlight down. "Why are you still on the rope? And why are you upside-down? This is not a vacation, Ranger."

J.W. jerked his legs violently until he was horizontal with the Earth then let a span of rope tear through the D-ring. He passed that ledge, though touched down on the next so hard, the weakened grommet on his chinstrap gave way. His steel pot ripped off and fell, striking ledge after ledge, clanking like the rifle as it tumbled into the valley.

J.W. yelped, "Off rappel," then sidestepped like a crab along the narrow shelf of rock toward the trees silhouetted in the meagre moonlight. While he sat waiting for the others, he leaned back against an ancient oak and lit a Chesterfield. After three puffs, he sank into a stupor.

Several bodies sailed past him, a couple of the men scraping their shins on the ledge. Their howling woke J.W., and he deduced he had finished his rappel on the wrong shelf. He groped back along the cold granite for the rope, but the cigarette he had lit to calm himself served only to blunt his night vision. After a fruitless search for a way down to the next ledge, he decided to climb back up through a wooded portion of the mountain toward the company and try again.

He struggled for twenty minutes, fighting the thick, sharp underbrush and tangled deadfall. His lips were kept moist by the trickle of blood from the laceration on his ear where his helmet's harsh, woven canvas chin strap had ripped a gash as it was ripped from his head.

The sustenance bequeathed by the few hours of sleep on the cattle truck was long squandered, and his climb in the utter black of the winter night became a ceaseless burn in his thighs. On the other hand, the graver the pain, the more evidence he was headed straight up. Partway along the ascent, his hips began moving more freely, and he was grateful for the improved level of conditioning from the Benning phase. A few meters farther up, though, it dawned on him the ease of movement was because his Swiss seat had come loose, lost on the mountain along with the D-rings. He stopped, debating the value of searching for the vanished gear, but accepted it was gone forever.

Following the flickering lanterns, he rose above the final ledge and came upon Gehring smoking cigarette after cigarette with his cadre buddies. "What *in the hell* are you doing up here? You were supposed to rendezvous with the rest of the company. Get your head out of your ass, Ranger. Wait a minute. No dumb ass Ranger's gonna fool me. You never rappelled down in the first place, did you? Answer me, Ranger." J.W. tried to respond but was gulping air so furiously from the climb that he could not form the words.

Gehring paced in circles again, mumbling, developing a more finely honed apoplexy with each rotation. "And where's your Swiss seat and your D rings? You destroy government property to avoid your mission, Ranger? I want twenty-six-and-one for being out of uniform, no helmet, and then you're gonna do twenty-and-one for lying that you already went down."

After serving his sentence, an additional twenty-five-and-one was awarded for having forgotten the twenty-and-one that came after the twenty-six-and-one. He borrowed a Swiss seat from one of the younger instructors and set off again on rappel. He spun rope through his gloves, aiming toward the glow of cigarettes and the sound of jabbering Rangers directly below. He managed to brake, landing head up on a ledge just above his cohorts. When he jumped the final six feet to join the main body, he lost another D-ring but spat on the rock wall, paying no heed to his debt of a regiment's worth of vanished combat equipment. He sneered at Gehring, who had appeared at the base of the mountain ahead of him.

It was deathly silent at base camp, aside from the rumble of deep snoring. First and Third Platoons had marched back hours before and were long asleep, bellies full. Fricker asked if he could go to the mess hall and check the oven to see if the mess sergeant had left anything for them.

Gehring groaned, "Oh, we got another wise ass here. Get into your rack, or I'll have you cleaning the ovens."

Second Platoon stumbled to their huts empty-handed and dropped onto bunks. J.W. dove for his bed and pulled a cold, rubber poncho over his head. He refused to look up when Gehring bristled into the hut, but the sergeant marched to J.W.'s bedside. "You got some paperwork to do for losing your helmet and that other equipment. You figure I'm gonna do it for you? No way, Ranger. Get yourself over to headquarters and sign for that stuff."

Triplicate, thin paper forms were waiting, each demanding an account of how the loss had occurred, details of the negligence, what steps he had taken to retrieve the lost matériel, and how the blunder would, not could, be avoided in the future. At the bottom was a note that his pay would be docked for the missing items. J.W. signed the last form and stood to leave, but Gehring called out, "CO wants to see you."

At the command shack, a light-skinned black man sat behind the olive drab field desk. Rather than the acidic expression tainting the majority of the officers he had met at Ranger School, this man's eyes were thoughtful. J.W. relaxed for a millisecond, but when he noticed the water-stained portrait of LBJ hanging behind the officer's desk, he was jolted back to reality. The man pointed to a chair, and as J.W. lowered into it, his arms tensed to deflect the first blow.

The man spoke quietly. "Relax, Ranger. I'm Major Powell, the Mountain Phase commanding officer. I got a call from Captain Vock. He says you took a swing at him, and I want to..."

J.W.'s jaw began to drop in disbelief, but it hurt too much to open. He muttered angrily through clenched teeth, "Excuse me, sir? That's bullshit. Excuse my language, sir."

"What's bullshit?"

"The whole thing, sir. The truth is he told me to have a seat in a shack just like this, then he slapped my face, and then he cold-cocked me in the jaw. You gonna do that to me too, sir?" J.W. pointed to his swollen, black and blue mandible. "This isn't from a bramble bush, sir."

"No, I am not going to touch you. But are you calling him a liar, Ranger?"

"Yes, sir, I guess I am."

"Are you saying you never hit him back?"

"That is correct, sir. Never lifted a hand. Never raised my voice, never threatened. Nothing."

"Huh. This is a problem, Ranger. If what you say is true, that kind of behavior went out with General Patton twenty-five years ago, and even then it nearly cost him his career. You are not to conclude that the United States Army tolerates that sort of conduct." He stopped and stared at J.W. "Lieutenant, you have an obligation to maintain the standards of the officer corps. You need to report this."

"I know I should, sir, but it's a captain's word, a Viet Nam returnee, against a Ranger's. No contest. And it'll cost me my career. You don't rat on people. Anyway, I'm still alive, sir."

"Weathersby, tell you what's going to happen here. I'm a fair man. I won't *make* you report this. I'm sorry to say you may have a point about making a statement with no witnesses. And I respect your attitude. Look, you're starting in the mountains with a clean slate. What happens between you and Vock when you get back to Harmony Church is your business. You can make a decision then about how far you want to pursue the matter. I will make a private note of what you told me. Man to man, Ranger, if you need my statement when you get back to Harmony Church, I'll get it to you overnight." He stopped again and sighed. "Just so happens, I believe you. But you keep this conversation between us, understood?"

"Yes, sir."

"Good. On the other hand, don't screw up here, or I'll have your ass on his doorstep before the sun comes up. Am I understood?"

"Yes, sir."

Powell scrunched up his face in a questioning expression. "You sound like you're from the City."

"Yes, sir. The Bronx."

"Huh. Me, too." Powell tossed him a banana. "Heard you got back too late to eat."

There were no pull-up bars at the mess hall in Dahlonega, but as at Benning, the shack was an ancient, peeling, wooden affair, and the line for breakfast at zero-four-hundred hours snaked around the outside through deep, black puddles of freezing mud. Many of the men on line sported clean fatigues, round faces, and hair more than a sixteenth-of-an-inch long. J.W. noticed that they wore rank as they pushed their way to the front of the line, ignoring the Rangers as if they were invisible. When Branch shoved one of them back, the soldier, a sergeant, cocked his fist, but his buddies laughed and pulled him away. J.W. took a step forward, coming to rest beside Branch. He tightened his lips threateningly at the sergeant, but Branch put his arm up against J.W.'s chest. J.W. glared into the man's eyes. Something was amiss, but it was too dark to tell just what. After growled threats and discussions regarding the sexual preferences of their mothers, the two sides drifted apart.

There was more to eat in the mountains: sausages and SOS—shit-on-a-shingle, creamed chipped beef on toast, the army's gastronomic showpiece. The largest tub held puddles of mashed potatoes.

Morelotto beamed as he loaded his tray. "I'm gonna put a few of those pounds back on. Watch me now."

Branch looked up. "What'ch you grinnin' at fool? Calories gonna burn off fast climbing mountains, faster than you can shove 'em down your throat. You ain't gonna make it. I'm tellin' ya."

J.W. started to open his mouth to add a comment, but his jaw was too sore, so he went back to his meal, the untidy sucking of warm milk through a straw. He tried the scrambled eggs, but they were too hard.

The initial tactical problem commenced that morning after the perfunctory visual rectal exams, with the three shrunken platoons marching from the cabins into the deep, verdant hills of the Smokies. The first miles were magical, the magnificence of the winter mountains carrying them on a wave of beauty. Even the fifty-pound packs seemed light, just part of the adventure into the clefts and warps of America's heart. For a few moments, J.W. believed they might have been the first souls in history to lay eyes upon some of the primitive rock formations.

They came to a remote parking area, boarded cattle trucks for a long drive, then marched a muddy, dirt road to a clearing. They were told to take a smoke break, the cadres urging them to light up two or three if they had them. Instead, J.W. smoked one and crawled into the fetal position until Fricker woke him and pointed wordlessly to the base of a very steep mountain crested with a patch of gleaming white.

Gehring addressed the company. "Gentlemen, this is Yonah Mountain. The foothills are behind you. So are your childhoods. We need to warm you up for the Mountain Phase, so you only have to run to that first ridge." He pointed to an outcropping of rock so far up the mountain, it might as well have been the snowy summit. "This is," he went on, "the first authentic tactical mission of Ranger School. We're keeping score for real. The clock starts now. Meet you on the ridge, those few of you who make it."

Morelotto laughed snidely, "Yeah, yeah, yeah, I've heard that shit before."

J.W. nodded in agreement, unwilling to move his mouth in comment.

The troops were marched along the road in a spread-out column, each man many meters from the soldier to his sides. Gehring shouted, "Left face...start the climb!"

Initially, the slope was gradual, and J.W. remarked aloud, "This is a cake walk, man. That shit at Benning was a hell of a lot harder."

Morelotto bobbed his head in agreement. But the grade abruptly steepened to ten degrees and, fifty meters farther up, to twenty. J.W.'s thighs and calves burned as if wrapped in glowing sheets of furnaced metal. His mouth opened despite the jaw pain, and he spewed a string of vituperative expletives that surprised even him.

A quarter of a mile up the mountain, under heavily laden packs, several troops dropped to the ground hollering in pain. Some cried they couldn't suffer another second of the torment, and Gehring ran to stand over them and laugh. Gangs of cadres, spotting Gehring feasting, rushed to the site to hover about single, prostrate Rangers, so many wolves over a kill. There were buzzed words, hissed words, and some grunted, but the chorus was the same: "DRIVE ON, RANGER."

Despite the shouted demands to restart, a critical mass of the beaten had fallen to earth as if crushed by the plague. Several of the infected managed to crawl to their feet, but instead of going on, they bounced down the hill faster than they had sped through the machine gun course. A mélange of their spent voices blended into a dissident chorus of, "I quit this shit!"

J.W.'s heart was drawn toward the pack of malcontents, magnetized by an overwhelming urge to turn and run. No one could withstand this level of misery, and he saw clearly that the cadres had gone too far. If every Ranger quit, there could be no recrimination. The man next to him surrendered and took off down the hill, and then a man on the other side of Morelotto. J.W. could feel the toxic organism of surrender entering him and then circulating, taking hold, and he lost a bit more control. It was okay, he told himself, the school was about to collapse.

Though he had begun to believe in the theory that Rangers did not quit over three more push-ups, that bravado was now surging out of his heart by the gallon. He managed two more short paces upward, but his thighs burned so, he dropped into a hunker behind a grand, old-growth cedar. When his legs loosened a bit, he thought about going on, but just three steps later, he gave up and returned to the tree. It was over. He wasn't even sure he could gather the strength to join the deserters.

What shocked him the most was that it had happened so fast, his final failure piercing him as if an unseen arrow. How could there have been so little warning? He sat against the tree, surprised and saddened, but not sufficiently humiliated to change his mind.

As the echoes of malcontents faded toward the cattle trucks, J.W. sat steeped in the silence and loneliness of the Chattahoochee National Forest. The rough, gnarled bark of the ancient cedar pinched his back, though it was less painful than the notion of resuming the trek, either up or down. He remained slouched and motionless for many minutes.

With the respite, he marshaled sufficient vigor to look up into the rocky palisades of northern Georgia and then down into the distant glens. He stared at the remnants of forest camps built by hordes of Civilian Conservation Corps workers, the legendary CCC, during the Great Depression. Though he began to hear the voices that had plagued him at Harmony Church, this time he surmised they were the ghosts of the souls who had sweat away months and years of their lives in these mountains, the dispossessed from the dying cities willing to work for just food and shelter. He thought of them laboring in the woods, never knowing what was to come next in their lives, when it would all be over, if ever, and when they would again hold their children and regain control of their lives. He wondered how many of those men could have imagined in their saddest dreams that they would soon die, not of starvation, but as victims of the wars in Asia and Europe to which they would be shipped as so much GI scrap.

With a bit of his strength slowly seeping back, J.W. smiled wryly. While it was hard to imagine the war into which he was soon to be hurled, in the deep hollows of his mind, J.W. Weathersby knew well *his* future was devoid, at least, of concern about sufficient food, enough clothing, or comfortable shelter. Only for the next hours would he starve, for soon he, too, would run down the mountain screaming in terror, "I quit this shit." He would sleep with his wife tomorrow night and eat at their little table in the apartment, devouring as many hot dogs and ladles of steaming beans as he wanted, as slowly as he wanted; a few days later, though it would be on the airliner to Viet Nam, he would sleep and eat and not be tortured or punched in the face. It gave him the strength to come to his feet and face downhill.

As he took the first painful step, his eyes caught the CCC camp again, and a voice yelled out to him, "You really *are* a maggot, aren't you? You are a parasite existing on our leavings and our sweat. We saved these forests and this country for you, and then we marched off to die, but first, we had to fight in mud and freezing forests and stinking jungles. *We* didn't cry when our legs hurt. You live off the fat of the land we built and do nothing but snivel about how you've been victimized."

J.W. boomed, "Living off the fat of the land? I have to go and fight in a war, no different than you. How are you any better? I'm doing my part. Leave me the hell alone, goddamnit."

The violence of his screaming further drained him, and he dropped back to lean against the old cedar, his eyes closing lest he see his own tears drip from his face.

The rejoining shriek did not surprise him. "Get off your fat ass, Ranger. Take the next goddamn step and drive on, you weak pile of crap."

This time the voice was familiar, though the words cut more deeply than they had at Harmony Church. J.W. glared about for the source of torment. He was sure it was Sergeant Hartack; he recognized the voice, but the closest man was a Ranger several dozen meters off, thrashing toward the ridge.

Farther up the mountain, he spied Branch and Bearchild near the head of the pack. He avoided their stares as they watched the company disintegrate. When they locked on J.W. facing about and taking the first step down the mountain, Branch shook his head sadly but resumed his climb. Despite his companions' disgust, J.W. began to descend in earnest. With each step, however, the screaming in his ears pounded harder. "Drive On! Drive On! Drive On!"

He covered his ears with his fists. The shrieking, though, pierced more deeply, and he turned in circles, eyes probing the forest for his antagonist. Still, there was nothing save his tree five steps uphill. He ran to it, rifle barrel in hand, the stock held over his head, as ready to slam the mother fucker in the skull as he had been to bash the stag. The torment was going to stop—nothing else mattered.

But the other side of the tree was barren and silent, as were the voices. J.W. sucked in a deep breath and launched into a mad sprint, orbiting the trunk, a crazed dog chasing its tail, though no matter how fast his feet churned the frozen mulch, he could not capture a flicker of his tormentor.

"Screw 'em," he cried as he started back down, yet the moment his foot landed, the voices began anew, and J.W.'s brain flamed with such agitation, he ran back up the hill to the tree, his rifle held higher, fists so rigid, his fingers drained to the hue of the snow-peaked mountains. He would shatter whoever had taken it upon himself to brutalize an innocent. He loped another twenty yards uphill to a colossal oak. Not a human trace at that one, at the next, nor at the one twenty yards farther up.

The voices, though, had ceased, and his awareness refocused on his legs. He took a step down. The voices returned. He started up again. The voices quieted. He took two more steps up. Silence. Five more, then ten, and he broke into a run, legs driven by a new force, a power outside his control. The pain and the voices degenerated into distant echoes. Two hundred meters farther up, nearly blinded by his rage, he tripped on a log and slammed face down. His jaw shot with such pain, it extinguished some of the fury, which allowed the hurt in his legs to flame anew. The voices resurrected. He'd fool them by standing and turning toward the ridge. He smirked as silence blanketed him, though it lived for only for a moment. As the first reverberation pierced through his violent breathing, he spun about and began to climb, though only in a lope. The volume of the voices waned, but not by much, so he pushed harder until they were a whisper. He was closing on the lead elements, men now barely able to walk. He yelped profanity after profanity, screeching for the assholes in front of him to clear the way.

When he again caught site of Branch and Bearchild, his heart seared with a passion fiercer than the flame in his legs, and he blew past them, throwing elbows, squealing about maggots and faggots, and cresting the ridge ahead of them. He looked about—he had arrived first.

He sucked air ferociously into blistering lungs but managed to focus down into the lower Smokies, at the Ranger camp miles in the distance, thousands of feet below, and at the tiny CCC camp. He was aware of neither his legs nor his jaw, conscious only of the absence of the voices.

⊷ ⊶

At the hot meal that night, J.W. dribbled cool split pea soup behind his bottom lip and worked it back slowly along his cheek, letting gravity draw it into his throat. Some of the watery broth trickled into his windpipe, triggering a paroxysm of coughing that punished his jaw. It was worse than if he had chewed. Branch's eyes opened wide, and he asked, "You okay, man? You need some help?"

J.W. shook his head and tried the mashed potatoes but couldn't open his mouth wide enough to suck in the lumps. As he puréed the soggy carrots, time ran out.

In the barracks, a lane grader, a former graduate of the Ranger School, sat on the end of J.W.'s rack and briefed them on the next day's mission. J.W. noticed the man's West Point ring and asked, "Lieutenant, do you happen to know Harold Steele? He's a good friend of mine."

"Yes, I do, Ranger. He just got married a couple of weeks ago. It was at West Point. Wish I could have been there. I was supposed to be one of his saber bearers, but my leave got cancelled at the last minute. Didn't you hear?"

"What's that, sir?"

"Shit, you didn't. Harold got it first week in country. Poked his head out of a tank and caught a VC RPG in the neck. Sorry to tell ya, man."

There was silence in the hut until the lieutenant drew a deep breath. "Well, gentlemen, we still have a lot of training to get done, so let's get…"

Fricker's hand shot up. "Lieutenant, sir, how long we gonna be in the field before we get back here? Can we take food? How long are we allowed to sleep in the mountains?"

The man answered quietly. "Ranger, make an effort to enjoy tonight. And, before you go to sleep, each of you make a field parachute by a tying a piece of this nylon cord I'm going to give you onto each of the four corners of your poncho, then tie the other ends to this hook I'm also going to give you. Questions?"

Morelotto grumbled, "I don't think that parachute's going to hold me. I want a big one like in jump school."

There was laughter until the lieutenant urged, "Gentlemen, the sooner you get the job done, the faster you get some rack time. When you're finished, put the ponchos in the wooden box by the mess hall." He gave them half a salute, mouthed, "Good night, men," and was gone.

⊷⊶

They slept for four hours, the 5 A.M., six-minute breakfast a distant memory by the time the company marched into the mountains. Forewarned that they were carrying their own C-rations, one box each, an allowance that would have to suffice for the five-day problem, Fricker griped, "That's only enough for half a meal. We're supposed to get double rations in the mountains, six boxes a day." His hand flew up.

Gehring grunted, "You ever heard of air-resupply, Ranger?"

Fricker shook his head, "Yeah, how do we know we can trust you?"

Gehring sighed, "Ranger Fricker, have we lied to you yet?"

Those who thought through Gehring's answer nodded in agreement that the cadres had not broken a single promise. J.W. relaxed, embracing the belief that Gehring had just made an inviolable covenant; the rumors were true, and they would receive sufficient food and sleep in the Mountain Phase, more than enough to make up for what was lost at Harmony Church. While it was comforting to hear, by late that afternoon, when the company stopped to chow down, most of the Rangers ate only a portion of their Cs, painfully hoarding in their rucksacks what they had the discipline not to devour.

J.W. cut little circles out of the cardboard boxes in which the Cs were packed, creating near-satisfactory covers for the opened cans he stashed in the bottom of his rucksack, a hiding place quite troublesome for another Ranger to infiltrate. When he tripped over deadfall or into pits, though, beef stew and fruit cocktail squirted out of the buried cans to soak into the pack. The Rangers who had copied his inspiration were transformed into trudging menus, wearing near-frozen stains of Ham and Limas, Pork and Beans, or Beef Stew.

By early the second day even the most disciplined Ranger's pantry was bare, the lot consumed during the sleepless night on the move. When the column stopped for eight-minute map checks, instead of eating food that hadn't been provided, they slept.

At noon the next day, Fricker challenged Gehring. "I thought you promised we were going to eat."

"I only asked if you had ever heard of air-resupply. Have you, Ranger?"

"Yes."

"Well, good, then you're the first to be picked for the air-resupply patrol."

Gehring took off through the forest selecting snoozing Rangers, creating a detachment to trek deeper into the mountains and rendezvous with "partisans," who would supply the next week's rations. J.W., having been unearthed napping under an enchanted blanket of dripping wet deadfall and moss, was added to the detail.

As Gehring threw a rock at the last sleeping man's head, a helicopter thundered into a landing zone fifty feet from the company. Without so much as a fare thee well, Gehring and the other cadres boarded, barely nodding to Poliak and several disembarking lane graders, brimming thermoses of coffee clutched under their arms.

J.W. assumed that the promised food was aboard the chopper, and he walked happily toward the ship until Poliak yelled over the rotor noise, "Where the hell you goin', Ranger? Just for that, you're on the list for the air-resupply. Get movin'."

J.W. snipped, "You can't hurt me. I already got chosen." He turned and swaggered toward Fricker.

By dusk, after crossing several mountains, the special air-resupply patrol established contact with the friendlies, but it had been a difficult day, one punctuated with repeated ambushes, many initiated by TNT explosions. The blasts deafened the Rangers, and they could not hear the belly laughs of cadres who had waited until one of their favorite students happened under the tree from which the explosive charge was suspended.

Surely, Sergeant Poliak opined, all of the unhappiness was a direct result of Ranger Platoon Leader Morelotto's decision to travel open trails and roads, defying the cadre's advice that he guide the patrol into the mountain forests. During one of the attacks, J.W. became bored shooting blanks and snuck off to rummage for food. He happened upon the flank of the enemy aggressors and captured one, a light-skinned black troop, a sergeant, whose eyes were just not right. J.W. finally realized one was brown, the other green.

Assuming it was the responsibility of a captor to abuse POWs, J.W. tied the man's hands behind him and began an interrogation. After J.W. dispensed with the perfunctory name, rank, serial number, and date of birth, he got down to business. "Prisoner, if you want to see your family again, you will guide us to your pack and provide us with C-rations. And by the way, what does your driver's license say in that little box marked eye color?"

The prisoner spat back, "What the fuck do you think it says? And you can eat me if you're hungry."

Sergeant Poliak, finishing a hoagie—bits of the salami, ham, a shred of green pepper, and a dollop of mayonnaise still on his lips and chin—heard the exchange and entered the clearing. Patiently, he smiled, "Okay, good job, Ranger, now let him go."

J.W., sensing a peculiar dearth of brutality in Poliak's order, answered, "No way, Sergeant. I got me a prize. I think I'll make him carry my rucksack and pistol belt."

"He ain't a slave, mullet. I said untie him."

When J.W. hesitated, Poliak threatened, "If you don't let him go, you will physically carry *him* on *your* goddamn back, along with *your* rucksack and *his* rucksack, for the rest of the Mountain Phase. You got that, you dumb fuck? You're not here to capture prizes, Ranger. You are here to suffer."

The patrol, sans POW, reached the partisan encampment at 2100 hours. It was pitch black and the students were soaked with super cooled rain that froze the instant it struck them.

In lieu of food, the partisan commander, a teenaged GI dressed in a beret and French peasant garb, proffered a thin apology that he read from a three-by-five card. "Rangers, we are truly sorry that our cache of rations has been raided by the communist enemy. Every burger, every bag of crisp French fries smothered in ketchup, every single deep fried onion ring, every bottle of beer, all of it has been captured.

"Your orders have been amended by supreme, I mean *the* supreme command." The nineteen-year-old pointed to a clearing on a wilted map of the Smokies. "You are to proceed to this point and prepare for an airborne drop of supplies."

J.W. and his mates dropped dejectedly to the dirt, unable to stop cursing, their eyes reddening, but there was little time to carp, at least in the sitting position, for Cadet Squad Leader Morelotto had them on a forced march, promising them a substantial cut of the rations once they were in hand.

At a clearing near the peak of a snowy cliff, a familiar voice called from the forest, challenging them for a password. J.W. shined his flashlight into the woods, lighting the face of the man with multi-colored eyes.

The soldier threatened, "You get that light outta my face, or I'll kick your butt from here to Thursday."

Then an older, even more familiar voice, interceded. "I'll take care of it, Sergeant."

J.W. glanced to his side. Weak rays of moonlight had peeked through the rain clouds, their glow reflecting off the mountain snow onto a pair of spit-shined boots. The timbre of the voice made J.W.'s shrunken stomach contract. A fern touched him on the shoulder. The voice spoke softly, "Guess what, Ranger?"

"Yes, Sergeant Cowsen, but Sergeant, I just don't understand what's going on. I thought you were only at the Benning Phase. You seem to show up a lot around here."

"Ranger, I get the feeling you are not happy to see me. I'm hurt. Never mind, you are officially the detail leader for this problem, and the next problem, too, and maybe the one after that if you don't hush up."

J.W. took the manila envelope and coordinated a march over forested hills to the drop zone, a clearing into which the food would be released from a flight of several army aircraft, or so it was written in the marching orders. J.W. was excited to be involved in a challenge involving aviation. He was sure that was why he had been chosen for the mission, the cadres acknowledging the complexity of the task. J.W. would be dealing with pilots, the army's most highly trained and intelligent personnel. His hunger and fatigue ebbed.

Over nearly impassable trails through rhododendron and laurel thickets, J.W. shaped a night free of ambush, but not of the threats of his men, who besieged him to revert to the roads, impotent aggressor attacks be damned. At midnight, the detachment reached the drop zone, the DZ, a large clearing sufficiently high to be covered in a thin patina of powdery snow. A lane grader briefed him that the friendly airfield from whence their supplies would be flown was to the south, almost certainly Fort Benning, he nodded to himself, and to assume that to be the direction from which the drop would come.

The lane grader then warned J.W., "Ranger, you only got ten minutes left. You got a lotta' work to do."

J.W. assigned men to wave flashlights when the plane first appeared, his scheme to alert the pilots. Then he selected troops to use their lights to form an arrow pointing to the center of the DZ and appointed a contingent to track the food-laden parachutes as they landed.

J.W. mustered the patrol. He warned they would have to gather the dropped rations and be off in minutes if they were to avoid attack by aggressor troops, for intelligence reports described enemy marines in the area who often let drops take place then attacked the Rangers and stripped them of their precious cargo. Vacating the drop zone with no delay was fine with J.W; the sooner they were on the road, the sooner they would pilfer the supplies and take an extra portion, and then another, for their trouble. He bore not a trace of guilt about stealing from Rangers who had stayed behind, wallowing in their time to sleep. Any one of those guys in the rear guard would have done the same.

Around 0200 hours, long after the slated time, the drone of a single DeHavilland Otter Caribou trembled in the northern sky. J.W. raced around the clearing, stumbling over sleeping men, desperately attempting to produce the mirror image of the drop zone he had squandered so many calories designing. When he had found enough of his men to form a facsimile of what his orders specified, J.W. tapped the barrel of his rifle three times with his bayonet, and those still awake flashed their lights into the night sky.

As the Otter flew directly overhead, saliva pooled in J.W.'s mouth, and his eyes strained to see the supplies falling toward his men. But the plane passed over the clearing without tossing out a single bundle, at least one he could detect. When the aircraft flew off south, J.W. used his radio to call the food-tracking patrol, but they were hysterical, reporting there had been no drop.

The pilot radioed he couldn't identify the target zone, and J.W. screamed, "Asshole," into the air, aiming a faultless bird toward the departing plane. Calming himself, he called the pilot, begging for another try, braying that he, too, was an army aviator.

The plane circled for a second pass, which now came, without warning, from the south. This time, glorious black bags hurtled to Earth. The first quasi-parachute landed next to Bearchild, and while the payload didn't strike him, the chute caught on his shoulder, ripping the sleeve of his fatigues. Several Rangers abandoned their posts to pounce on the cardboard C-ration boxes, ignoring both Bearchild and the missiles splashing and pounding around them.

As Fricker came to one of the packages, he lifted it by its parachute and yelled out, "Hey, it's our ponchos. They're back!"

J.W. would have ripped the top of a C-ration can off with his bare teeth had there been a single can of Cs. Inside the cartons, instead, were bandoleers of machine gun ammo, smoke grenades, white phosphorous, "willy peter" signaling grenades, and fake Claymore mines containing not a speck of C-4 to heat the Cs that hadn't been dropped. There were also dozens of cans of M-1 ammo, bullets which did not fit their M-14s. Those cans, all the ammo cans, gave off deep thuds as they were flung into the bushes by men seeking to avoid carting the extra weight back across the mountains.

Morelotto hobbled in circles, bewildered, but his ears suddenly perked like a Doberman pinscher's. The plane was circling back. He shoved troops out of his path as the next consignment dropped in a streak. He reached the first package just as it bounced several feet back into the air. On its way down for the second time, it struck his hand, whacking his middle finger into a dislocated zig-zag. He screamed in agony. The lane grader, who had scrutinized the chaos without expression, walked to Morelotto, grabbed the crooked digit by the tip, then yanked and torqued until the joints cracked back into position.

Morelotto yelped but recovered when he eyed the offending parcel at his feet. With his arm and middle finger held protectively above his head, he dove onto the package. "Cs! Finally!" he yelled deliriously, slashing with his good hand at the cardboard with a bayonet. When the thick carton pulled apart, however, Morelotto paused for a moment then swore at the top of his lungs, "Shit! It's filled with rocks."

Several of the Rangers pushed past Morelotto, sprinting after the Otter, though J.W. called them back. He cupped his ears to listen, but the plane did not turn back, and as the drone faded, he tried in vain to raise the pilot on the radio.

The men drifted about the clearing ripping the last of the crates apart, finding only ammo and bags of scarlet earth that had apparently been flown up from Benning, lest the Rangers forget Harmony Church. When it was clear there was no food, case after case of bullets and rocks arched into the forest to be stored, the Rangers calculated, forever.

As the last of the supply drop was hurled with mounting violence into the brush, a heavy night shadow meandered into the drop zone. Fricker blurted, "Sergeant Cowsen, nice to see you, Sergeant."

"Gentlemen," Cowsen declared placidly, "you will retrieve the ammunition from the area in which you have stockpiled it. We expect you to transport every piece of ordinance back to your unit. You must not lose a single round. Your lives depend upon finding each of the thirty-six cans and boxes you've cached. Your leader here, Ranger Weathersby, has a record in his orders of every box that fell as manna from heaven. And his tenure in this school depends upon your searching abilities."

J.W. peered at them anxiously as the men turned slowly toward the bushes. Cowsen shook his head and laughed, "And the rest of you will go through the school twice until you find that stuff. Now get your butts a movin'."

A dozen grumbling men dug OD cans of ammunition out of the shrubs then presented seventeen of the metal containers to the lane grader, who was warming coffee over a can of Sterno. He did not count the cans. He did not even look up, but sent the patrol back into the forest where it took until zero four hundred hours to find the last can, the one that had rolled into an old CCC latrine.

J.W. divided the freight amongst the troops, Bearchild stepping up first, smiling and nodding in appreciation as J.W. handed him a share. No one noticed when Bearchild drifted off into the woods and emptied most of his load in the mud, replacing it with a scaffolding of dry twigs and leaves upon which he placed a single length of linked bullets. J.W. also drew cans of machine gun ammo, but promptly tossed the bandoleers back into the brush and carried only the empty cases, panting and stooping under the burden each time a lane grader shot a glance at him.

The contingent marched toward a crest high in the Smokies, J.W. still in command. He headed for the ridge, despite the fact that it was several miles out of the way, mumbling that if his men had to walk along the side of a mountain, below the ridge but parallel to it, they would trek for miles and hours with one foot higher than the other. "Look what's gonna happen. Your ankles'll be burnin', and you'll blame me. So, we go straight up the hill, get the pain over with in a single spasm, like at Yonah, then hump the ridge where the walkin's easy. Anyway, the aggressors are too lazy to climb up there."

Rocky Mount, jewel of the range, loomed directly overhead as dawn broke. With the light, command of the detail passed to another Ranger, instantly leaving J.W. devoid of the adrenaline that had driven him through the night. For the first time since taking over as commander, the hunger and fatigue burned inside of him more intensely than anything he had imagined at Harmony Church. Even the mountain's magnificence could not blunt the torment. He quit moving and dropped onto the frozen earth, grumbling about giving up until Branch slogged by breathing like a horse. He grabbed J.W.'s fatigue jacket and screamed, "Move your honkey ass."

J.W. crawled a few feet then rose and finished the climb. At the peak, he looked out at the ancient forests and the towering granite. In one of the valleys was the highway that had deposited them in Dahlonega. J.W. followed it to the bend in the road where his troop truck had hit the bear, and he remembered looking up at the very peak on which he now stood. How far away it had seemed that first day, and how sure J.W. had been even Ranger cadres would never force men to climb that high.

The Ranger leader read his orders aloud. They were to reestablish contact with the partisans, and this time there would be food—it said so right there in the orders. They pushed on for hours, drawn along only by hope. Nearing the rendezvous point, the lead scout discovered evidence of an ambush. He doubled back, empty ammo cans clanking like so many cowbells, and recommended to the Ranger platoon leader a roundabout route to join the partisans from the north and avoid the trap. That stratagem, though superficially sound, neglected the snag that there was no way to signal the friendlies that the patrol of Rangers had altered their route and would approach from their rear.

Bearchild and J.W., assigned point, crawled an inch at a time until they stumbled upon a partisan lying on his back in the dull shadows of dawn, snoring as though a logger cleaving the forest primeval. The partisans were dressed in the peasant garb of the French underground, a throwback to the glory days of the Second World War. J.W. laughed to himself, imagining the third sub-basement of the Pentagon where an anonymous brigadier general sat, dawdling his last army days away as Director of the Ranger School Planning Committee. Weathersby could picture the man, feet propped on his desk, reliving his youth as a shavetail lieutenant, conjuring scenarios to aggrandize the French underground. Perhaps he recalled the most captivating night of his life and the magnificent young woman with whom he had made love and cried. And though they could barely understand the other's tongue, they held each other that entire night as they slept in the forest of Provence, the rockets and tracers sailing high above their passion. The general surely remembered that dawn, the woman's warm, soft, nude body draped over his, her weeping as they rose and dressed. In minutes they would leave each other, drifting separate ways, never again to know the intensity of that night.

The tribute seemed very silly to J.W., and the allure of sex even more absurd than the garb of the partisans. But if those troops were the bearers of food, it mattered not if they were dressed as Barbie Dolls or flower children. J.W. prepared himself to treat them with respect and to grovel at their feet, to beg, to crawl for hundreds of meters along rock-hard crooked roots, if that is what it took to establish contact and curry sufficient favor to warrant a lump of congealed fat.

As he approached the snoozing partisan, one of J.W.'s ammo cans caught in a piece of twine stretched between two trees. While he had seen the twine before crawling under it, and even thought about it for a split second, in his state of conscious stupor, J.W. ignored it as simply a piece of string in the forest tied neatly between two trees. To puzzle out how that fiber had found its way to an altitude of three thousand feet in the Smoky Mountains would have cost far too many calories, and he had gone on his way, crawling toward the fantasy of a bite to eat. On the other hand, perhaps the energy spent in processing a thought would have prevented the ignition of the flare that was attached to the end of the string, and the subsequent explosion of light that blinded him and shocked the partisans awake.

The Frenchmen took up battle positions, one screaming, "Blow the Claymores."

The man sleeping on rear guard jumped behind a tree, and Bearchild lost sight of him. Not sure if the Claymores were real or blanks, Bearchild crawled to J.W., grabbed him by the fatigue shirt, and tried to pull him back. J.W. stood and turned to run, but as he took the first step, a body pounced on his back. His jaw struck the ground first, the pain a dagger in his head. An instant later, it was his arm that throbbed.

"What the hell are you two doing here?" a voice on top of him cursed as he twisted J.W.'s shoulder further into a crushing, numbing hammerlock.

"Sir, we're Rangers. We were supposed to make contact with you for food. We need to eat real bad," J.W. pled.

J.W. screwed his head to the side. He could see his captor's uniform. It was, indeed, a partisan, thank God, but when he stretched his neck to get a better look, he noted the most salient feature of the man's persona—rainbow eyes.

"If you were really Rangers," the man hissed, not recognizing them, "you would have entered from the south. You're lying. You two don't know the password, do you?" J.W. was silent. "You don't even look like Rangers. Look at you. Your uniforms, if you can call them that, are ragged. And you got bad attitudes."

"Bad attitude, sir?" J.W. whined as the hammerlock tightened.

"Yeah, all you want is food. You don't give a damn about us, do you? You just want to feed your fat gut, that's all. Did you bring the ammo for us? No ammo, no food."

J.W. looked toward Bearchild, who twitched in a subtle nod. "Yes, sir, we have the ammo, sir," Bearchild offered confidently. "May we show you, sir?"

Bearchild sat up and untied two cans of .30 caliber, belted machine gun ammunition from his rucksack. He opened the tops but aimed the contents away from the soldier and toward the early rays of sun. The single layer of brass shells glinted for a single moment in the partisan's eyes, and then the cases were closed before the man could see the strands of grass poking up around the bullets. It brought to mind the tales returnees spun about the slicky boys on the streets of Saigon pushing up sleeves for a tenth of a second to flaunt a dozen bogus Seiko watches. The partisan loosened his grip on J.W.'s arm and ordered the two Rangers to their feet.

J.W. pulled his cap over his brow as far as he could and looked down, staring at the hoarfrost. Another man came out of the perimeter carrying blindfolds. As he started to tie one over J.W.'s face, the guy with the funny eyes burped in laughter, "No shit. Lookey who we got here. This is going to be a fun week. I think I'll have this asshole carry *my* pack for the next few days."

The two captives, each dragging several rucksacks, were led into the partisan camp. A marvelous aroma floated through the clearing, and J.W. peeked under his blindfold to witness four men preparing Cs. He waited to be fed as a condition of the Geneva Accords, but not one of them offered so much as an empty can of Ham and Limas to lick out.

J.W. and Bearchild were tied to a tree. The blindfold was pulled from J.W., and the sergeant with the strange eyes began an interrogation. "So, fuck face, how many men in your patrol?"

"None, sir."

"Cut the shit. I'm going to kill you and dump your body off a ledge. Is that what you want?"

"No, sir."

"I'm just gonna kill you right now and tell 'em your pal here did it." The sergeant took an antipersonnel grenade off his web belt and very carefully pulled the pin, but kept his hand tight on the spoon. "How many men do you have with you?" he threatened, placing the grenade into J.W.'s crotch, but J.W. remained silent. "I'll ask you one last time, numb nuts, how many men?"

As J.W. was about to blurt, "Twelve, sir," Bearchild, who had worked his hands free, grabbed the sergeant by the neck and throttled him violently. The grenade fell to the ground, and Bearchild dropped to his knees, though as he did, he realized there was not time to heave it away. Instead, he spun around and put his body in front of J.W. and the sergeant.

Other than the din of crashing branches and cursing Rangers, there was silence—no explosion, no carnage. The sergeant howled, "Get the hell away from me. The grenade's a dud, like your friend here."

The main party of Rangers swept across the tiny camp, subduing the partisans and pilfering Cs and cigarettes. A lane grader caught up with them, however, and warned, "If one Camel is missing, the whole company will pay with three extra weeks in the Mountain Phase. Don't believe I can do it? Watch me now." With that counsel, booty dropped from armpits and the shorts of those Rangers who still wore them.

A shaky truce was struck when the five partisans were released, and the Ranger leader was brought to a small clearing where a tarp covered a pile of burlap bags. The partisans agreed to trade their food for ammo, but only after the leader demanded, "American soldiers, show us the bullets you bring."

Bearchild was the first to step forward and place his two cans back on the ground for Act II. Droning a dissident, Native American chant and whooping occasionally, he put his fingers in the latch of the ammo cans. When he came to a crescendo in his mantra, the tops sprang open. With all eyes on Bearchild's face, he swung the lids closed. He stood, bowed respectfully, lowered the volume of his prayers, lifted the ammo cans, and quietly walked to the back of the congregation to stand and stare at the sky while he mumbled a prayer.

One of the partisans looked at Bearchild curiously, strolled toward him, stood there for a second, and screamed, "Hey, Ranger, what's your problem?"

Gillette rushed to the partisan's side. "Sir, Ranger Bearchild's presenting his gift in the traditional Native American fashion. He feels true pride in having helped his friends, and you see, now he's stepping out of the limelight in humility. He will not answer you. He can't. If he did, he would lose face in front of his totem. We need to respect his cultural imperative."

The partisan leader shook his head, muttered, "What the...?" then nodded impolitely and went back to his companions. J.W. was ordered to open his own cans. They were as empty as they had been the moment he'd tossed away the ammo nearly a day before. He was told to stand by a tree. One by one, the others opened their cans to angry stares. A new final tally of the air drop episode was reckoned: twelve foodless men; one of them, a peculiarly behaved soldier, was allowed to continue the march without an ammo can; and eleven other Rangers slogging OD canisters stuffed with wet dirt. Added to their burden were several burlap bags of rotting carrots and reeking potatoes. The grand prize, however, was presented to J.W. Weathersby by the sergeant with multi-colored eyes. J.W. was ordered to tote a bamboo cage stuffed with live, cackling chickens for the march across the mountains back to his platoon.

Morelotto, still holding his hand and middle finger in a defensive posture from the drop zone dislocation, demanded to know where the real food was hidden. That apparently disturbed the partisans, who suddenly spoke with harsh, idiotic accents, claiming that what the Rangers had been given was better than their own families were eating; their children were starving because they had gathered these supplies for their liberators.

Morelotto apologized. "Hey, man, I'm sorry, I didn't know. Okay?"

Liberators? J.W. was already borderline psychotic, and that word triggered another hallucination. He was back in France again. It was early 1945, and he was suffering from battle fatigue. In his dream, he walked along the cobblestone roads of Provence, searching for the gorgeous Gaelic women who, he had heard, poured from bunkers to welcome their liberators. He dreamed of sweeping one of them up in his arms, perhaps like the general at the Pentagon, and taking her home to love for all his life. His new wife would cook food every day: French food, German food, even English food. They would eat and eat and eat and stop only to sleep, the notion of making love never entering his fantasy. He pictured the marriage ceremony and the reception and then the unending feasts, but just as he sat down to pheasant under glass, and *de cheval* smothered in *pommes frites,* a huge explosion made hamburger of his dream.

J.W.'s eyes opened, not to lace tablecloths and a cornucopia of victuals, but to muddy trails, ferns, frozen moss, and the same Rangers with whom he had tracked the hills of Georgia for weeks. There were no aquiline-featured women, no Paris, just a bang as one of the partisans brought him back to Dahlonega with a rock to the forehead. The stone cracked so hard, he was thrown into yet another fantasy. This time, he found himself on a pockmarked, Vietnamese road, dodging the ancient French mines that laced the countryside. But just as he focused on the tortured face of a Vietnamese father carrying a dead son, a kid who had been playing with a leftover grenade, more trauma befell him. A volley of fists pounded his head, and J.W. regained consciousness.

"Put that cage of chickens on your back and bring it to the main body of your unit, asshole."

J.W. looked up into a green eye. The man's knuckles were still tensed, and J.W. flew into a rage. He vaulted to his feet, quickly shoving the partisan and taunting, "Come on you funny-eyed fuck. Hit me when I'm on my feet if you have the balls."

J.W.'s body stiffened in a fury he had seldom known in his twenty-three years. He would fight back this time; nothing would stop him. But as he cocked his arm, the other partisans came to their colleague's side. J.W. looked anxiously about for his brothers, but the Rangers had seized the opportunity of preoccupied partisans to busy themselves scouring the camp for wayward Cs and smokes.

With the absence of friendly backing, J.W.'s bravado cooled a degree, and then another, and he took a step back. "Just let me catch your ass out there in the woods, pal."

⊨⊨ ⊨⊨

The bamboo rungs of the chicken coop assigned to J.W. had sharp ends which tormented the open sores on his neck, leading him to stuff his shirt with a wreath of frosty moss; but the ice melted and dripped in unending rivulets down the inside of his fatigues. After several hours J.W. gave up, finally ignoring the problem of his bleeding neck, and forced himself to trudge the hills numbly, without commentary, toting his miserable, feathered companions.

Somewhere in the forest, Cowsen appeared, pulling up alongside Fricker. "Ranger, you know how to peel carrots?"

"Yes, First Sergeant."

"Good."

"Ranger Branch, y'all went to basic training, didn't you?"

"Yes, First Sergeant."

"Y'all learn how to peel potatoes on K.P.?"

"Yes, First Sergeant."

"Good. Every man peels what he carries."

When it dawned on J.W. that even Cowsen would not eat a live chicken, and that someone was going to be tapped to peel J.W.'s passengers, he went to Cowsen on a break. "You know, First Sergeant Cowsen, I don't kill things. I didn't grow up like that."

"Just take it easy. No one's gonna make you eat those chickens, Ranger," Cowsen assured.

All he could think about were his neighbors in New York returning from their annual, autumn, up-state hunting trip with the carcass of a buck strapped across the right front fender of the father's '56, Borough of the Bronx, Department of Inspections (For Official Use Only) Ford. Every year, they roped their upside down, tongue-dangling trophy to a branch of the backyard oak tree. When J.W.'s father asked the dad why they had strung it up, he answered, "To let it bleed," allowing his eyes to roll up over J.W.'s father's lack of masculinity. The corpse hung there for what J.W. remembered as weeks, rocking from the same branch that held the family swing during July and August, the one on which their eight children had courted mates on balmy summer nights. It wasn't until that moment that J.W. realized the beast had been strung up, not to render it kosher, but in a display of urban machismo. The food that wasn't in his stomach churned in disgust.

Cowsen stopped them late in the afternoon. "Rangers, I'm gittin' a mite hungry, aren't you? I am going to instruct you in the military method of making chicken stew."

The contingent of malnourished gathered at lightning speed, all ears and drool. "Ranger," Cowsen called to J.W., "y'all hand me one of your ammo cans. Transfer the mud into your rucksack, right at the top so I can check it every few miles. You know, I'm still very upset that you destroyed military equipment. It hurts me inside. Knock off fifty-and-one for your trouble. Maybe I'll feel better after that."

The eleven men creakily joined him, and J.W. smiled that this time the watershed had undeniably been crossed. A true core of men had merged. They were a detachment working together smoothly, an elite military unit so seasoned and well-oiled, not one compatriot would be lost from that moment forward. J.W. vowed to carry every straggler on his back, to claw the earth to save a single Ranger from the ignominy of defeat by the enemy. The aggressors may have been Green Berets and troops from the 197th Light Infantry Brigade, but the *real* enemy was the corps of cadres. Sides had been chosen. There would be no more solicitous obedience; the game would be played by Ranger rules from then on. He called aloud at push-up fifty-one, "Glory halleluiah!"

Cowsen scraped the remaining mud out of J.W.'s ammo can with a stick, though spoonfuls still clung tenaciously to the inner walls.

He smiled, "There's gonna be a tad extra pepper in yo stew tonight looks like, Rangers."

Cowsen then stripped the black rubber gasket out of the shoe box-sized, olive drab, metal ammo can, adding, "Gentlemen, you will be poisoned if you cook in an ammo can with the gasket in place. May also blow up. Be pieces of chicken stew and Ranger splattered all over the Smokies."

He stared directly at J.W. "Ranger, bring me a chicken." Then he called over his shoulder toward Bearchild, "Make me a fire, but don't burn down the whole doggone country while you're doin' it."

Bearchild commenced gathering twigs while Weathersby stared into the grey sky. J.W.'s brain plummeted into another of the surrealistic Ranger trances in which his past roiled up, memories of long forgotten incidents sent back to guide him in making sense of the present. This time the convoluted string of ideas began with the chickens. J.W. had carried the caged poultry on his shoulders for hours, and it hadn't been an unsoiled journey. Early in the march, J.W. squandered dozens of precious calories brushing slimy, pistachio-green chicken droppings from his fatigues. After a bit, he let the guano gel on his uniform and found it peeled far more readily in that state. Then he discovered that if he let the chicken turds freeze into a rigid tumor, he could flick away the solid crescent with a snap of his finger. That left only the slightest residue of the silvery-gray ellipse on his fatigue shirt, barely a trace on his fingernail, and cost several tenths of a calorie less than making a sweeping motion with his hand. As the volume of droppings grew, however, he became impatient, and reverted to wiping away the debris while still soggy. That left a checkerboard of tapering, green, mucoid smears which froze anyway, more quickly, and spread the cold over a greater surface area of his fatigues.

For hours, J.W. trudged, passing the time attempting to remember a formula or a theory relating to the transfer of heat from a lump of chicken crap to cotton fatigues. It was probably taught on one of those Saturdays he missed ME 153, his mechanical engineering course in thermodynamics. His college guidance counselor, between puffs of the pack of Camels he'd polished off during the fifteen-minute session, had enrolled J.W. in thermo for the fall semester of his sophomore year. That was during football season, when J.W. was fighting to gain the attention of the varsity coaches.

The course instructor was a young graduate student, not quite five-foot-four, who had scarcely begun to shave. He did, though, pack two K&E slide rules slung in hip holsters. He made it clear from the first session, eyes fluttering in disdain, that he had no sympathy for those with interests other than the natural sciences. As Captain Vock despised married men, the thermo instructor especially detested crew cut football players.

The thermo class was subjected to a quiz every Saturday morning at 8 A.M. J.W. was usually with the football team at that time, a thousand miles from both campus and thermodynamics. The instructor, while allowing no excuses, was kind enough to drop the highest and lowest quiz grades at the end of the semester, averaging the remaining. The rub was that J.W.'s maximum score for the exams he managed to attend was a fourteen, and that was out of one-hundred points. That and one of his three zeros were dropped.

A "B" in the course was one of the highest grades he received in college, and he was surprised, but silent, when the report card arrived. While his average had been a nine, more consequentially, the instructor had been caught as a founding member of a ring of scientists dropping pennies into hydrochloric acid to reduce them to the size of dimes. These coins fit nicely into the pay phone, the washing machine, the drier, and the candy and soda machines on the dormitory floor. This was a transgression for which the numismatist and his cronies had been swiftly expelled from graduate school, and everyone in ME 153 was awarded a "B" just to keep them quiet.

Cowsen hollered, bringing J.W. back to life, "What the heck you babblin' 'bout, young Ranger? I said, carry me a chicken over here."

The bamboo cage was lashed shut with wet twine, the knots locked so tightly, the only way to open the cage was to cut the hemp. J.W. drew his bayonet across the rope a bit too eagerly, the tip cutting through several of the bamboo rungs. Two chickens scampered from the cage. Weathersby leapt forward to seize one by the neck, but the animal raked J.W.'s hands with its feces-covered talons, so he let go.

Fricker pounced on the other bird and shoved it back in the cage. J.W.'s prey was gone in a blur of dirty feathers. It came to rest teetering on the edge of a sheer mountain wall. Bearchild, who was returning from the forest with a harvest of logs, saw the errant bird, ditched his load, and ran to the precipice. He dropped into a crouch ten yards from the chicken and waddled silently along the brink. As he extended his arms to snatch dinner back, the thing tumbled over the rock face, fluttering and cackling, its wings shedding feathers as it bounced off ledge after ledge. A dozen seconds later, it was but a speck of white splattered on the valley floor a quarter-of-a-mile below. It took a while for the report of a dull thud and weak cluck to reach them.

After the push-ups, Cowsen asked if J.W. would be kind enough to bring him one of the remaining pullets. The sergeant waited, eyes to the heavens, until J.W. was standing over him with a bird. The first sergeant howled, "Ring its neck, Ranger."

"Can't do that, Sergeant," J.W. protested quietly. "Remember, I don't kill things."

"Ring its neck, or I'll ring yours!" It wasn't that J.W. didn't eat meat. J.W. loved chicken. But this *was* a chicken, and J.W. had never touched a live one, or for that matter, given a single thought to where fried chicken came from. But he was a Ranger now, and he tightened his hands around the bird's neck and squeezed, ignoring the thrashing and clawing.

Cowsen tolerated the bungled execution for a few seconds then brayed, "No, no, Ranger, gimme the doggone thing." He grasped J.W.'s chicken by the head and twirled until the bones cracked. Then he cut through the neck with a pocketknife and let the body drop to the ground, its scrawny legs running the headless carcass in circles. Everybody laughed, and soon there was a chicken, or at least a piece of one, in every ammo can. Cowsen added unpeeled carrots and the mushy, blackened potatoes Morelotto had lugged. He poured in putrid water, latched the top back onto the can, and threw the whole thing into the fire Fricker was minding for Bearchild, who was dragging a trunk of a deadfall oak into the clearing.

The meal was undercooked and squishy, with rare, semi-defeathered, lukewarm chicken parts, a bloody ooze running from between the layers of the pink meat. There were lumps of crunchy mud suspended in the broth, and flakes of chewy, olive drab paint. Subsequent field manuals directed the cans be lined with tin foil before cooking, for the paint had been deemed toxic to humans. Nonetheless, J.W. sat back against a huge fir and enjoyed the most delicious of meals. He savored each drop of stock and used his lips to make mush of the unwashed potato skin and sour carrots.

He wrote the recipe on the back cover of his Ranger Manual, though when he tried several times over the years to reproduce that stew, it was never again quite that delicious.

The total energy intake, however, came to only a fraction of the double rations they had been promised. In two days, every morsel was gone, the final tally amounting to a couple of carrots, half of a potato, and a shred of chicken per man per day—that is, for those who didn't scarf the lot on the first sitting.

By week's end, J.W. was able to open his mouth half-an-inch, but if he tried to drop his lower teeth just a bit farther, an electric shock raced through the left side of his head. He limited his intake to water, but that was all they were able to find anyway.

Over the next weeks, Class 68-B crawled the entire range of mountains, often coming upon the long-abandoned CCC camps J.W.

had seen from Yonah Mountain. Although he searched the ruins for scraps of food left behind by aggressors or hunters, there was never a particle. He sought, on one of the longer marches, to estimate the number of calories Rangers had gambled away over the millennium in the futile pursuit of "free" food. Back to thermo, he wondered why the uncounted units of wasted heat had failed to warm the Earth.

Survival, he discovered, was a full time endeavor, one that demanded a total commitment to foraging. And one had to know, *de novo*, what was edible, for alpine gastronomy was not one of the courses in the Ranger curriculum. Fricker learned that lesson when he ate berries from the last autumn's crop, shriveled little black specks that had managed to stay on the vine through the winter winds and snows. The taste in his mouth after he spit them out lasted for days, and he asked a lane grader, "Hey, sir, how long before all I can taste is my usual didn't-brush-my-teeth odor?"

The lieutenant answered, "Hey, hunter-gatherer, how would I know? I use a toothbrush three times a day."

Light snow began to powder out of a gray afternoon sky. J.W., despite being clothed in only shreds of what was left of his fatigue pants, and even with the building numbness in his fingers, stared deep into the majesty of the Smoky Mountains and became once again enthralled. As dusk fell, with snow blanketing the forest, he experienced a sense of elation that surprised him, and which made thoughts of the frigid hours ahead tolerable. J.W. could already taste the bitter cold and was aware of how much worse it would become before Sylvester reappeared. Maybe tomorrow would creep up into the low thirties.

He bitched to Gillette, "It's gonna be hell tonight, man. Wish we had some real clothes."

Gillette laughed, "Remember, a mountain is a poor man's overcoat."

Morelotto overheard the metaphor and growled, "What the fuck are you talkin' about now?"

Gillette began a symposium, but they were rousted to their feet. As they moved up the first hill, Morelotto became so hot he took off his fatigue jacket and draped it over his rucksack to dry.

The temperature crept steadily downward, slowly at first, then precipitously. Hours passed on the move, and the patrol neared the ridge of a particularly high range. There, the feared cold met them head on, and at the first break, Morelotto went to the patrol leader and ranted, "Man, we gotta keep movin'. You gotta keep your men warm. Don't you know the mountains are a poor man's overcoat?"

J.W. entered a thicket of trees, found a tiny spot in which to crouch, and prepared to sleep. It was too cold, and his sweaty fatigues were beginning to freeze. He gave up and lit a cigarette. As he settled against a tree, his hand brushed a soft lump in the dry snow. J.W. hesitated, remaining perfectly still, calculating if it was worth the labor of reaching forward a few inches to discover what he had unearthed this time. Back and forth his mind raced, a struggle between conserving his strength on the one hand, and the possibility of hitting upon treasure on the other. He accepted that chances of stumbling upon riches twice in one Ranger School were diminishingly small, and he castigated himself for having squandered the mental energy to consider the impossible. Eventually, he rebuked himself for the abject waste of treasured calories, thinking about whether he should be thinking.

J.W. Weathersby's newfound discipline failed seconds later, and he found himself contemplating his options carefully, considering the mathematical odds of striking the mother vein of happiness in the barren wilderness of the frozen Smokies. By the time he reminded himself that the process of deliberation was costing calories, he could feel himself warming with the effort he was putting into ignoring the lump in the snow. "It's enough, goddamn it! Drop it."

Fricker called out, "You okay, Weathersby?"

"Never been better," he called then whispered, "asshole."

But it clung there, the gnawing ache, the festering stomach ulcer, like Cowsen's presence. J.W. heard the lead elements begin to crunch iced-over puddles. He moaned that if he didn't grasp the moment, the lapse would loiter on his mind like a song that wouldn't go away. The pressure built until J.W. summoned the intestinal fortitude to let his fingers open from their clenched curl and reach forward, believing that if he did so slowly, he would consume fewer units of energy. His numbed hand traveled through the snowy powder, inching along, until his fingers brushed against a foot-long mass. He sucked in a deep breath of glacial air, mumbling an entreaty to the Lord, then examined, by Braille, the lump of hope. The object softened subtly in the meager warmth of his hand, and J.W. soon appreciated a fur-fringed, leather tube with smaller tubes protruding from one end. It was as if an animal had died and turned inside out. J.W. slid his frozen hand a millimeter further into the bushy part. It fit like a glove.

He pulled the bequest out of the snow and played his flashlight on it. It was a glove, a thick, leather, fur-lined, arctic glove dropped by an aggressor or a cadre, and not that long before; the leather was still dry.

J.W. considered screaming for joy to let the world know of his good fortune, for he was in sole possession of what had been, seconds before, only an angry emotion for the being who'd lost it. He congratulated himself. But for his courage in reaching forward, the treasure would have remained there, a forgotten, useless bump on the great expanse of the Smokies. J.W. reveled that perhaps he had served his penance, that life had taken a positive turn; and even if his discovery did not portend heavenly dispensation, at least his right hand would be beautifully cozy and warm for the rest of his life, and even better, for the rest of the night.

"No," J.W. talked to himself aloud, "I'm not broadcasting my good fortune to the world. Not this time, no way. First thing you know, the less fortunate's gonna demand to borrow it. Then they'll let it slip through their fingers, lose it just like the dud who dropped it in the first place. Then again, I should whisper to the patrol leader that the enemy is close by. But if I do that, Cowsen'll want to know how he knows, and then I'll be fingered, and then, since it belonged to the enemy, I'm gonna have to give it back. No way. I'd rather get ambushed."

For the next several miles, J.W. switched the glove from hand to hand until the column came to rest in a large clearing. As the delay approached thirty seconds, J.W. made the decision to invest the energy in squatting and sleeping until they started forward. If the respite lasted only two minutes, he would cherish it as an eternity embellished with a single hand that was not numb.

Anywhere else on Earth, he thought, two minutes was an irritation at a red light, perhaps the wait for a McDonald's burger, but in the hand he'd been dealt, it was a major breakpoint, for any stop beyond that meant calamity for the platoon leader, generally a relief of command for having become irreparably lost, and then another fifteen minutes while the new platoon leader woke up and oriented himself. Yet J.W. felt no guilt in greedily savoring the momentary reprieve purchased at the price of the student leader's angst. If time was money in the real world, time was beautiful, warm, cozy sleep in his.

Absorbing what heat he could from the mulch, J.W. ignored his other carnal concerns and slept, unaware of the movement around him. The cold intensified, but he didn't care; he simply crawled into a tighter ball. J.W. had heard that dying in the frozen mountains was peaceful, pleasant, like drowning, once one let go of his fear.

He accepted that the only spark left in him was not the beauty of the Smokies, but the warmth of his right hand. Surprised that the heat began spreading throughout his body, his eyes closed tighter, but soon he understood that the warmth had only brought his brain to the point that he could recognize his life ebbing away.

Even with that awareness, he could not force himself to his feet, nor did he wish to. He fell deeper into the semiconscious sleep, unable to move and never intending to again. Slowly, his heart filled with a smile, as he believed his debt for the transgressions about which his mother had accused him had been paid in full. He drifted further and further from Dahlonega, though he did not know where he was destined. The journey proceeded slowly, a sensation of great peace soon blanketing him. He was approaching nirvana, all the threats he had endured dissolving in the warmth of his hand.

Just as he felt himself ascending into paradise, his spirit suffered a sudden jolt. The quiet and the peace, the caress of completion, all of it was jerked away as J.W. absorbed the thud of a combat boot in his chest. He looked up to see First Sergeant Cowsen waving an icicle, readying to christen a new platoon leader.

"Are you out of your cotton pickin' mind? A Ranger doesn't sleep in the snow. Haven't we taught you anything?"

Cowsen walked on. He looked about the clearing until he found Bearchild and tapped him on the shoulder with the frozen spike. Bearchild, who had been asleep as well, jumped to his feet, grabbed the manila envelope, and shouted, "Yes, First Sergeant." He studied the orders.

Cowsen drawled easily, "Figure out where you are. Come up with a patrol order. We need to get moving. No delays. It's getting cold. I'm countin' on you."

Bearchild paused. He, too, had not consulted a map in days, perhaps weeks, but Cowsen walked away, leaving Bearchild to solve the problem without hovering over him and spewing invectives.

Bearchild was instantly awake. He placed a dirt-blackened, lacerated index finger on the symbol of a high ridge deep in the mountains. Cowsen walked back and looked over Bearchild's shoulder. His expression did not change. Bearchild declared to the student staff he'd gathered with a crook of his finger that the platoon would descend from their ridge and plow through the thick underbrush to hit the enemy in their rear. He knew the aggressors expected them to follow the ridge, and that that would have been the most tactically expedient route, but he pointed to a funnel in the topography, the perfect setting to ambush the Rangers.

"Gentlemen, every one of us is consumed by hunger and sleeplessness. We are devoid of the stamina to pick through the thick stands of rhododendron encircling the mountain slopes, but like Ranger Gillette showed us, we can and we will."

Cowsen asked Bearchild, "Ranger, you sure you want to forsake the comfort of this here ridge for that tangled forest?"

Bearchild did not answer. He called Branch and J.W. aside, arguing there was no way to move quickly through the dense underbrush if the men remained attached to rucksacks and pistol belts. Bearchild asked, "What 'a we do to move faster?"

Branch suggested they leave their gear behind, but J.W. warned that if they did, Cowsen would inform the aggressors, and the Rangers would never see their equipment again. "They let us get away with it once, but they won't again."

Bearchild thought for a second longer then proposed, "You and Branch stay behind. Guard the stuff. Only ones I can trust."

Branch and Weathersby appeared conjoined twins, identical expressions that made them look ten years younger; a glow radiated as if they had won Irish Sweepstakes. Their hands and faces would be spared the violent ripping and tearing of those forced to navigate the rhododendron thickets on the attack, and then run back into them to chase the enemy in mop up operations. Though the two would not have the prospect of pilfering Cs from the vanquished, both agreed with a wink and a nod that their fortunes had just rocketed. They swore several times to Bearchild that they would take their guardianship seriously.

After Bearchild briefed the platoon, Cowsen announced that two Rangers from the previous class had been lost at the base of that very mountain. He suggested they keep an eye out for the bodies. An extra meal would be granted the first Ranger to uncover the remains. J.W. suddenly reconsidered and advised Bearchild that perhaps he ought to go along on the attack, but Bearchild ignored him.

While the other Rangers dropped their gear by squads and loaded M-14s with blanks, Branch and J.W. studied the plane geometry of the three rows of rucksacks, making a show of neatening the baggage to insure fields of fire. Finally, they blackened their faces with camouflage sticks.

Bearchild came up to them and shook his head. "All I asked you to do is watch the gear, not win World War Three. Don't go to sleep, you two." He paused and eyed them carefully. "Don't let me down."

Branch and J.W. nodded solemnly to him and then to each other.

As the patrol left the clearing, Branch turned to J.W. "We got lots of time. No one's comin' up here to bother us."

J.W. squinted at him. "My guess is that Cowsen'll radio the aggressors and tell 'em to capture us. That guy with the funny eyes wants to kick my ass. They'll take all the gear, and that'll make Bearchild look bad. You remember that guy who lost all our stuff?" J.W. looked up into the gray sky. "Ah, LaVoy, that was it. That'll piss off Bearchild, and then we'll have dissension in the ranks, and Cowsen'll be delighted, and I'll be standing before The Man by dawn. We gotta be careful."

Branch laughed scornfully, "*You* be careful. We just got a gift. No way Cowsen's gonna allow us into direct, unchaperoned confrontation with the aggressors, man. Last thing he needs is another dead Ranger."

"Yeah, but what about finding those bodies? An extra meal," J.W. groused.

"That's bullshit," Branch hissed. "Cowsen's bluffin'. Even if somebody finds the bodies, there ain't gonna be no meal. You think Cowsen's carryin' an extra ration in his pack for a Ranger? You nuts or something, man?"

When the last elements of the patrol departed, Branch and J.W. picked out a broad, fine-barked tree and sat, shoulder to shoulder. J.W. looked at the extra battery they had been given to power their radio transmitter, just in case. The radio was the platoon's lifeline, and the two security guards' main function was to use it to direct the platoon back to the clearing after the attack.

J.W. spoke quietly. "The two of us are gonna spend the next coupla hours on our butts. How does that make you feel?"

"What are you, a psychiatrist or somethin'."

"No, man, I'm just sayin', we're gonna be cold as hell out here, not movin' and all, and I sorta had an idea."

"I'm listenin'."

"Okay, we sleep until we hear the shooting. We wait a few minutes and check in with Bearchild, ask him if he needs us. He'll say, 'No,' but he knows we're awake, and so does Cowsen. Then we use the Ranger Bird, mostly to get them back here. That way we save the juice in the battery already in the radio, and I use this one to heat us up.

Within thirty seconds, the two had snapped their ponchos together into a makeshift shelter. They gathered moss, made a soft seat on the ground, and leaned against their packs.

Branch cackled, "What are we smokin'?"

J.W. suggested, "Your Chesterfields."

They lit up but didn't inhale. Instead, they puffed like emphysemics, stoking the glowing cigarettes to warm the inside of their primitive sanctuary. When Branch finally complained that his eyes couldn't take the smoke, J.W. took the cigar-box-sized spare radio battery out of his rucksack. "Here's what a university-trained engineer can do. Watch closely."

J.W. cut the cardboard cover away. With his teeth, he stripped the rubber insulation from the wires of the six cells and twisted the bare copper together, creating a dead short-circuit in the battery.

Branch coughed. "Shit, man, that stinks."

"Yeah, but you can feel the heat, right? Tell the truth."

As the fumes of sulfuric acid hissing from the boiling battery became more than J.W. could bear, he, too, started coughing. When the battery's metallic casing melted and a drop of molten lead fell onto J.W.'s bare crotch, he tossed the contraption out of the lean-to.

J.W. grumbled, "Okay, it's gone. Happy now?" Branch was silent. J.W. looked at his companion to see that he had collapsed backwards against the tree, lifeless from the poisonous vapors. J.W. raised the edge of the poncho and waved his hand madly to move air toward Branch.

Branch's head lolled forward. "Man, that shit's freezin'. Close the gotdamn window, fool." There was a pause as J.W. tucked the edges of the poncho back under his boots. "Man, why does it take so long for the white man to get their women in bed?" Branch tipped an inch closer to J.W., opened his eyes, and asked, "Is it true?"

"Is what true?"

"Do white girls really make you wait 'till the third date before they put out?" When J.W. hesitated, Branch snapped, "Well?"

"Okay, I gotta admit something, if you promise you won't say anything."

"No, man, of course not," Branch reassured.

"See, I haven't been all that lucky. I mean, well..."

There was a period of silence, for so demanding a conversation was calorically quite expensive, especially to argue a subject so impossibly remote. Branch faded for nearly five minutes, and J.W. was frightened he'd really passed, but the little man gathered himself. "That's the way the white man lives. Why y'all so 'fraida everything? What the hell they gonna do to you, Ranger? Yell? Who gives a shit? It's just a puff of meaningless air. I been yelled at all my life 'cause I'm black. Teachers, cops, you name it, man. And I'm here same as you, cold and pissed off. Yes or no?"

The last volley sapped a good portion of the steam left in Branch, and he fell back against the tree before muttering, "They told me I wouldn't get an M.A. in psychology 'cause I was gonna do my thesis on black officers in the Army. Touchy shit. I wrote that black troops don't respect black officers. They don't believe a black man can lead 'em. I can't take that shit, Ranger. I can't live with it."

Branch leaned forward again. He crushed out a cigarette angrily. "Man, I'm sad. No, I'm pissed," he slurred his indignant words. "Do you know the shit heaped on my head since I was a kid comin' up in Bed-Sty? Guns, and fightin', and hatred. Then my mother looked around one day and said the place was a pit, enough was enough, so she moved us uptown to Harlem where there was supposed to be community, people who saw a brighter future for the black man. There was a renaissance of the spirit, she said. Blacks were moving up in the white man's world. Blacks from Harlem on TV, right there on the screen working alongside the white man. And it wasn't Amos and Andy any more. This was the black man and white man laborin' for the same boss, both takin' the same shit.

"She showed me articles about there bein' no more segregation in the military. And black athletes bein' respected. Jackie Robinson got a college degree from UCLA. Scholar athlete. He comes to visit Harlem but don't go near Bed-Sty, which is right in Brooklyn where the Dodgers live. What's that about? Honest, Ranger, I was scared that I wasn't good enough to live where there were smart people.

"I didn't want to move, but off we go to Harlem, Miss Bessie Branch and her only living child. Wasn't no different. Maybe less death by guns, but the same death by despair. Schools falling apart, teachers angry and scared. We laughed at them. Those teachers were pussies, do-gooders. Wanted to be part of the 'black experience.'" Branch waved his head like Stevie Wonder, with a sickly smile. "Bullshit. Still so much fightin', and drinkin', and knowin' that's all there's ever gonna be. And these liberal, asshole teachers from the burbs. So brave doin' it in the ghetto. Arrogant bastards. Thought they were so gotdamn smart. Bring the white man's ways to people who ain't never gonna have the dough to live those dreams. Who wants to, anyway? And I'm just as gotdamn smart as they are. I told 'em that, too."

Branch took a long breath to go on, though J.W. interjected, "Bet you did."

He was silent again, but not for long. "Proved it, too, on the SATs and the GREs. But you know what's worst of all? No matter how many degrees or how much rank I ever get, man, the best I'm ever gonna be is the token darkie. That's if I don't get uppity. If I open my mouth in the real world, honkey'll be on my ass like stink on shit.

"Lot of 'em yelled at me." He leaned forward and spit. "Lots of pissed off officers when I was doing my thesis. Man, Infantry Branch made me report to this major general at the Pentagon—fifth subbasement or somethin'—to get permission to go on with my research."

J.W. interrupted. "Guy have a picture of a French lady on his desk?"

"What the hell you talkin' 'bout? So Second Lieutenant Branch spent fifteen minutes doin' Step and Fetchit, beggin' the sucker to let me do my work. Mutha fucka threatened me with the loss of my commission and a bad conduct discharge if I didn't drop the subject. Shit, I didn't have the packing varnish polished off my butter bar, and I'm about to lose it.

"I left there and had a nervous breakdown right on the street. Don't you go tellin' anybody, but I called a cab and checked myself into the hospital. I was so scared, I was shakin'. I heard the ER doctor say 'The Negro in Three's a drunk. Send him up to the detox floor.' I got dressed and left."

He lit another cigarette. "Bad conduct discharge? You know what that means? That's the end of me. All the work out the window. Kill my mother, all she did for me.

"When I got the courage to tell her, she said, 'Son, I raised you up to be knowin' right from wrong. I didn't spend those years on my knees washin' floors to be sendin' you to no college to have some white man in a cave take it all away. I'll find you a senator or, if I have to, the president. You're worth it.' Two weeks later, Miss Bessie Branch gets a letter from a Southern senator sayin' he'd take care of the matter. He musta, 'cause I never heard another word from the army.

"Worked my butt off and got my gotdamn M.A. last year from Penn, man, Ivy League. Published my thesis in the *Journal of Society and Race.*

"And here I am dyin' just like you. Don't matter who yells at ya, Ranger. No man can do a damn thing to you unless you let 'em." Branch fell back against his rucksack.

J.W. reflected on his own life. "I guess I wasn't raised so brave as you; wish I had been. I know you're right, man. I just don't have the balls to do it your way. I spent my whole life tryin' to please somebody else, not 'cause it makes me feel good, but 'cause if I don't, they'll get angry and won't love me, I guess. I don't know."

"And who the fuck is *they*?" Branch nagged.

"I don't know that either."

"That's the whole point. I'll tell you who 'they' are, Ranger." He shot forward, fully awake, and as enraged as J.W. had ever seen him. "It's a creation of your own frontal lobe, man, reinforced by coaches and teachers and parents and friends' parents and all the rest of 'them'." He spit hard onto the dirt. "It's everybody who's protecting their own little slice of the pie. Got nothin' to do with who *you* are; it's about upsetting their little social apple cart. You don't do it their way, you're threatening the only thing they got—their conviction that their meaningless lives mean something. Puttin' you down makes 'em feel like they contributed to society, that they were morally appropriate, and God'll take care of them in the next life. They never lived through shit like this, so they got to make up something to hang on to.

"See, if you're a little different, and you are, and you don't do it their way," he started to lift his hands to make quote marks, but exhaled and shook his head almost imperceptibly, "their way, well, then you're challenging their state of grace. If they're wrong, and been wrong for so long, they're fucked in terms of salvation. They may think about it for a minute, but they'll go right back to way they were. Everybody does. They're right, period.

"Only way they got to get back at you is for all the scared people to join together and shame you. So some fat, meaningless, piece of shit points a finger at you and declares, 'You're in trouble, boy'. You stiffen up, turn red, and they win. You let 'em win. They can't really do anything to you, but you believe they can, so you lose.

"System works, man—look at you. I'm right and I know it, and everyone else's nuts, man. You been living by the mirror image of that. I'm tellin' you a fact. Look, man, you ever hurt anybody?"

J.W. thought for a minute. "I don't know. I've done some stupid shit in my life. Threw a fire cracker in a swimming pool. Some of the paint peeled off."

"I don't wanna hear 'bout no silly kid stuff, man. I mean, you ever *hurt* a man? Physically attack somebody? Put a gun in his face, rob his ass at gunpoint? Burn down his house, even harm him with words?"

"No way."

"Well, then why haven't you done it your own way, with your head high up in the air? If'n you ain't hurtin' nobody, what do you care if you do it a different way?" He was thundering, and his eyes moistened. "Tell me that!"

J.W. contemplated those words, wondering if it was too late to change. Branch sucked in a very deep breath and shook his head sadly. "Mirror image, I'm tellin' ya." He lit yet another cigarette and leaned back against the tree. "I'm not sayin' I got it all figured out, but me and Bearchild were talkin'. We're the tokens, and they're watchin' us real close. Every mistake I make, I can feel it burning in my heart, Ranger. If we don't do good here, we're disappointing a lot more than him and me. I feel like Sisyphus pushin' the gotdamn rock uphill. Everywhere I go, it's the same. I can't stand it anymore. I know what I said, but you got it easier. Wears on ya. I'm gettin' very close to the end of my rope with this shit. One of these days I'm gonna let 'em all know. May cost me my place here, but I'll just go back to school. Don't make no never mind. Pen's mightier'n a Ranger Tab."

J.W. gasped, "No, man, you'll get it done. You gotta stay. I won't make it without you."

The warmth had built in the shelter, and Branch's face relaxed. By the light of the cigarette ember, J.W. saw Branch smile. "Look, man, you'll do fine. Just remember what I said."

They leaned back and fell asleep, shoulder to shoulder.

A moment later, J.W. looked up to a snippet of light under the edge of the poncho, the burning ember of the cigarette that had just rolled out of Branch's hand. The radiance, though, was yellow, not red, and J.W. shook Branch. "Get up, man. Platoon's comin' back. I can see flashlights."

J.W. lifted a corner of the poncho. A blast of light blinded him, and he shouted, "Get that thing outta my face, asshole."

It wasn't, though, Second Platoon. "Shit," Branch grumbled, "it's Sylvester. Shit, man, it's past dawn." He jumped up and surveyed the clearing. The equipment was gone, every stick of it. They listened for the rustle of the platoon, a sneeze, the Ranger Bird, but they caught only a light breeze rattling icy twigs of oak and white birch.

Sliding and slipping down the heavily wooded mountainside, they finally heard the timbre of human voices coming from the base of the hill. Branch mocked, "Assholes didn't get very far."

But when J.W. kept moving down the terrain, Branch motioned with his hand to slow the descent. J.W. shook his head and kept moving, calling over his shoulder, "I smell breakfast. Hurry up."

Branch whispered heatedly, "Slow down, damnit. I smell trouble."

J.W. turned back for an instant and laughed happily, "You smell trouble. What are you, Bearchild all of a sudden? Well, breakfast's a callin', pal, and from now on, I'm goin' after whatever the hell I want. Some guy taught me that last night."

J.W. laughed, consumed with the fantasy of filling his belly. He dragged forward, stumbling toward the scent of cooking C-rations. He spied the far corner of the camp—a dozen troops huddled around a gigantic bonfire. With field jacket hoods pulled over their heads, they were eating with spoons from OD tins. Though Branch motioned frantically, Weathersby continued forward, lost in the vapors of tinned Hot Dogs and Beans. His mind even blocked the image of the colorful unit patches on the shoulders of the field jackets around the fire, and the fact that Rangers had not been issued field jackets.

He stumbled on, a phantom in the *Night of the Living Dead*, step after quavering step, until a powerful tug on his fatigue blouse drew him to the ground. Branch hissed as he dragged him back into the deeper brush. "Honkey, what the *hell* you doin'? Man, that's the gotdamn enemy, fool."

J.W. turned back to the fire. His eyes opened as large as the tops of the C-ration cans tossed carelessly about the clearing.

"Shit, man, look at their uniforms. The suckers are clean. What's that unit insignia? Damn, First Infantry Division. That guy's a returnee, man. None of the Rangers are returnees, are they? Shit, we just coulda got killed."

The aggressor with the Big Red One on his shoulder turned to stare at the perimeter, his body tightening. Branch pushed J.W. farther into the deadfall, and the two lay perfectly still. The man stood and took a few steps toward the edge of the forest then turned back to his buddies. Without a spoken word, a Green Beret joined the infantryman, pacing cautiously five steps behind, covering him with an M-16 at the ready. They studied the crushed snow and followed the fresh tracks. Branch tensed to jump them, but the men paced back toward the fire. The Green Beret called out, "Watson, Dayton, cover us."

Branch sprang to his feet and smacked J.W. in the head. The two crashed through the rhododendron and Douglas fir for an hour, until the ruckus of loud voices drew them through the colorless morning. They approached slowly, scoping the residents of the next clearing before crawling into the bivouac area.

It was Second Platoon. Most of the Rangers sat back against packs, eyes closed, not stirring. Branch and J.W quickly took seats at the fringes, leaning back and closing their eyes.

Bearchild surveyed the edge of the platoon area and approached Branch and J.W. He thanked them for their good work. Even Cowsen stopped by, congratulating them on the security they had provided the main body. He then appointed Bearchild, Branch, and J.W. permanent rear patrol, their mission hence to hang a half mile behind and guard against sneak attack. J.W. thought he might have heard Cowsen mumble, "You're the only ones I can trust."

Branch could not hide his smile and whispered, "Another gift, man. Can you believe it? Okay, here's what we're gonna do. Listen to me. We rest when the platoon leaves. Stay right here. We don't move. They travel so slow, we can catch up with 'em in a minute. We do that every now and then, make a racket so they think we've been there the whole time. We give the platoon leader a report then melt back into the jungle. We do it again, takin' a longer rest period each time. Lord be lookin' out for us."

Bearchild stared questioningly at Branch. "You two sound pretty chipper for having been up for the past month."

Weathersby became more serious. "Look, my man, I got nothin' to be scared of anymore. I'm in control of my life from now on. No more torment. I got this thing figured out. You just watch me."

Eight minutes later, Second Platoon staggered off north. The three groaned theatrically but swaggered out of the clearing to the south, stopping after two hundred meters. When they could no longer hear the platoon, Bearchild and Branch dropped down against a tree and slept. J.W. took five minutes to search the clearing for cans of Cs not licked clean then joined his brothers.

Fifteen minutes later, Bearchild nudged their legs when the hubbub of the main contingent could be heard thrashing the woods less than a quarter of a mile away. The platoon was walking in circles, so they followed the cacophony for a while, caught up, straightened out the platoon leader, then dropped back into the woods. The next break was at the edge of a barely paved, two-lane highway that snaked through the hills. The three smoked and talked. Branch inquired if Indian women were prudes on the first date.

Bearchild grumbled, "Ask Weathersby."

As they rested by the side of the road, an antediluvian, battered, hand-painted milk truck came around a bend. J.W. began to flag the vehicle down, laughing, "Gents, anyone for chocolate milk?"

Branch grabbed him and pulled him into a ditch. "Fool, they're aggressors. Look at 'em. Clean, drinking coffee. It's a captured vehicle. They're trolling for stray Rangers to bag and torture. Don't move. And black people don't drink chocolate milk. Now you're goin' to tell me you do."

The three peered out from the snow-covered swale to see two men with scraggly beards, a driver sitting on the bare springs of a high seat, and a passenger standing, both in civvies, their hands laced around steaming mugs.

J.W. observed, "They're aggressors, and I'm Napoleon. Look at 'em. The driver doesn't even have teeth."

As J.W. started to crawl forward, Branch grabbed one of his arms and Bearchild the other. Branch whispered angrily, "You won't have any either if you move one more inch, Ranger. I'm not in the mood to screw around with you. Just sit here and don't budge. Can't you get it through your thick head that there's only two kinds of people in the world—bald and dressed in OD rags like us, and all the others. And don't forget all the others are your enemy, too."

J.W. yanked his arm away and yelled in a whisper, "What the hell is in your head, Ranger? A man looks different than you, and you get all hot and bothered? Why is it so frightening? That's the whole problem with society today. Segregation and all that. I say we take a chance and flag 'em down and beg for coffee and milk. Maybe they got some breakfast rolls on board."

The notion of coffee and donuts gave Branch pause, and he looked questioningly at Bearchild, who shook his head.

Branch spoke for his silent partner. "Be smart, man. You got enough paradise here, just bein' away from the main body. No one's botherin' us. Be cool. Don't blow it."

J.W. took a deep breath. "Yeah, I suppose you're right." He was quiet for a minute then laughed. "But, think about this. If they *are* hillbillies, we're supposed to look down on 'em like they're bone-heads, right?" His friends sort of nodded. "I mean, no education, no future. But guess what, *they* got freedom, those milkmen. Man, can you imagine buying a cup of coffee whenever you choose, and ridin' in some old, broken down truck instead of slogging through snow? That is satisfaction, my man. Those are some lucky sons 'a bitches."

J.W.'s brothers thought silently about the proposition then bobbed their heads solemnly.

After the rusting truck ground by, the three crawled out of their warren and crossed the highway. J.W. whined from the rear, "When I'm living in the real world again, no matter what the hell's goin' on in my life, when I open my eyes every day, I get a cup of coffee. I don't care what you two do. Nobody stops me. Never again, not in my life. You got that? And if I want to drink chocolate milk, I will—a quart if I want, every morning."

Bearchild asked, "Branch, what the hell's up his ass?"

At the far side of the highway, they followed a dirt road parallel to the forest path taken by the bulk of the platoon. They detoured when they spied a ramshackle farmhouse surrounded by a host of dilapidated, widely separated outbuildings, each more rotted than the last. They searched for food in the least eerie of the huts, and when that failed, they traversed a field stubbled with the remnants of last year's corn. Bearchild kicked snow away from the stalks, searching for an errant cob, something to boil in melted snow and drink, to warm and sweeten their mouths.

J.W. asked how he knew about such things, and Bearchild offered more words than J.W. had ever heard from him. "When I lived on the res in South Dakota, my mother could make a meal from nothing." He squatted and pulled hollow stalks from the earth then tossed them disgustedly into the wind. "This field's bare, picked clean. The poverty's in the Smokies, too. Got here long before the Rangers did. Just like home."

J.W. shook his head. "Bein' honest, Bearchild, I can't imagine not havin' food. Not in America."

Bearchild went on. "The fields at home were bare by late winter. Nothing in the house, hardly even firewood. The only thing makin' us warm when we went to bed was a cup of corn tea, old corncob boiled in water. Heat drew the sugar out. My mother served it to us children in chipped porcelain mugs. It was the best. I always saved a swallow or two by my bedside to give to my little brother, Steven War Bonnet. First thing in the morning, Steven'd get up and smile at me. But that's..." Bearchild's voice trailed off, and J.W. discerned a flicker of anger in his friend's eyes.

An uneasy silence settled over them, and J.W. walked toward the main house, drawn by the wisps of chimney smoke and sparks rising into the still, frigid air. "I'm goin' to the house to beg," he grunted.

Branch and Bearchild followed several meters behind, scanning left and right like commandos, until a few orange embers drifted toward them. One landed on Branch's bare forearm, and he let out a yelp, brushing the glowing fragment from his skin. It was just a fraction above the tiny keloid from the wasp at Harmony Church. The burn blistered instantly, and he cursed bitterly, piling the lesion with snow. The disquiet brought a face to the window of the decaying farmhouse. A moment later, the thick, rudimentary door creaked open, and a cachectic, sallow-complexioned man stepped over the threshold. He left the butt of his antique shotgun on the porch, but gripped the barrel, allowing it to lean subtly forward.

Branch was about to turn and walk away. He was holding a snowball on his forearm but kept his other hand tightened on the business end of his M-14, as if to use it as a club. J.W. held up his palm to stop Branch, whispering, "Let me handle this, man. I'm hungry. I'm gonna appeal to the man's sense of patriotism. This is America. We're all on the same side."

Branch hesitated but riveted his eyes on the silent figure in the doorway. He whispered back, "You mean you're gonna appeal to his sense of Bronx bull crap. Let's get back to the unit. I don't like this shit."

J.W. walked, hands in the air, toward the rickety porch. He stopped ten feet away and smiled respectfully. The old man's grip tightened on the barrel, and the stock lifted a fraction of an inch off the porch. J.W. looked into the dull eyes and began tentatively.

"Sir, we are three cadets from the United States Army Ranger School. I understand that some of the summer Rangers have been inconsiderate of your crops in the past, and we have been dispatched by the high command to assure you that will not happen again, sir."

The farmer stared at J.W. slack-jawed, as did Branch and Bearchild. When the farmer's hand relaxed on the barrel, J.W. continued. "Sir, I do have to tell you that we are very, very hungry. We haven't eaten in the week it took us to get here. We're very shaky, sir. I'll be honest, sir. We need some food, anything you got. We'll pay for it, of course. I mean, we are not here to beg, no sir, we're gonna pay."

The aboriginal face stared a while longer, but the shotgun butt lifted over the threshold and the man took a step backwards. J.W.

imagined the farmer was gaining distance to gun them down with a single blast, but the man leaned the barrel against the doorsill and stepped back out.

"Lemme' see de culluh o' y'alls' money."

The man's words introduced a snag into the convention of the deprived. While J.W. paused to decipher the quasi-English, he suddenly understood that it was now four worlds that had collided on a patch of frozen corn waste. The common thread was want, though the one with the bleakest life and dimmest future was the least needy.

The meeting began to reduce to a very simple formula: on one side of the equation were the four or five dollars the Rangers were promising, and on the other side of the equal sign a bowl of corn grits or three slices of stale bread. They could see in the man's face that, while money in those mountains was a windfall, he had survived for a very long time without it. He could go on—they couldn't.

When J.W. turned back and saw two pairs of rolling eyes, he concocted several ludicrous fairy tales, but the man's face had hardened, and Weathersby made a decision to tell the relative truth.

"Well, sir, as I'm sure you know, they don't let us carry any money. But, sir, I swear on my father's grave we're going to pay for any food you would be kind enough to share with us. We need the food, sir. We will send you the money, I swear."

The farmer paused, looked out at the pleading, desperate faces, then backed inside, keeping his eyes pasted to his guests. He took the gun with him but left the door ajar. The three shrugged, moved slowly forward, and craned their necks to scrutinize the innards of the house.

"It's mountain modern," J.W. murmured as they peeked at the bare walls, raw furniture, and an old TV with more snow on the screen than in the fields.

The man reappeared silently, catching the Rangers many steps closer to his home than where he'd left them. He paused and looked down at his shotgun, a gesture not lost on Branch, who slowly tightened the grip on his ammo-less M-14. Bearchild grabbed the muzzle away. The old man stared at Bearchild for a moment then turned away, disappearing into the house. The Rangers stepped backwards into their original, snowy tracks.

Hearing clacking and banging, Branch whispered, "Hey, man, I wanna leave while I still got all my parts." But when they realized the din was cooking pots, in Pavlovian reflex, the three crept closer to the shack.

The old man came to the doorway holding several packages wrapped in grease-stained, brown paper bags. He walked onto the bowed steps and cautiously handed the parcels to J.W., started back, then turned to them and drawled, "Y'all teh dem ta leave my crop 'lone, y'all hear? We ain't dun nuttin' to harm y'all."

J.W. was not sure if he saw a tiny smile as the old man turned back to disappear into his hovel.

The three rushed off with their haul, Bearchild stopping at the edge of the property to copy the name "Sullens" from the long-neglected mailbox. At the tree line, they halted. J.W. ripped at the paper sacks, revealing six guano-coated, raw chicken eggs, half a loaf of fresh bread, and, J.W. became light-headed when he saw it, a hunk of fatback bacon. They even had paper to start a fire.

J.W. suggested, "Okay, let's eat. I say start the banquet. Right here on this patch of forlorn dirt."

Branch shook his head. "For-fuckin'-lorn my ass. What the fuck are you talking about? Are you crazy? Take one bite, and then let's get into the woods and cook a proper meal like civilized human beings."

Bearchild nodded, took an egg, wiped off some of the chicken droppings, pulled the shell apart with his grit-blackened fingernails, and lifted it toward his mouth. Before a single strand of the slime spilled, he extended his neck, put the shell to his lips, and allowed a continuous string of yellow mucous to slide into his throat. His eyes rolled.

J.W. could not decide between the raw egg and the raw bacon until Branch ripped a piece of the salt-covered pork off with his teeth, chewed a few times, gagged, and spit the lump onto the ground. J.W. retched in parasympathetic sympathy and settled for a bite of bread but snatched Branch's expectorated calories from the dirt, slipping the chunk of lard into his pocket.

A gentle breeze was blowing toward the platoon, and the three sought a deeper corner of the forest, finding a secluded thicket where they dug a pit, then constructed an earthen oven. Bearchild started the fire with a single match, igniting the greasy paper that had wrapped the bacon. He was soon placing thin twigs directly on the oily, yellow flames, barking orders with snaps of his fingers for special-sized twigs, sticks, and, finally, logs.

In minutes they drooled beside a respectable wet-wood fire, with Branch and J.W. nodding in concert that it was time to start cooking. Bearchild, however, would not hear of such disrespect until the deadfall tree trunks he'd dragged from the forest had ignited. It became one of his crowning efforts, though J.W. voiced concern about the smoke.

Bearchild assured, "Fire's hot. Got a lot of energy. Smoke'll rise high above the platoon."

Branch stuck his head between them and turned to J.W. "Didn't you ever take thermodynamics, man?"

Branch sliced the bacon into bite-sized, one-inch cubes then cracked the eggs and dumped them into his pristine mess kit, at once a frying pan. Bits of shell sprayed in as well, though there was not a breath of protest. Branch thrust the loaded pan into the midst of a now spectacular blaze, gathered flattened stones, and placed them near the fire. He turned his eyes to Bearchild, who cut the bread into ragged slices with his bayonet and leaned them against the steaming rocks to make toast.

There was nothing left to do but wait. J.W. pulled a lump of bacon from the pan, but Branch hit his hand. "Man, a watched pot never boils. Put it back."

The three sat beneath the forest behemoths, cozy and warm, enjoying a fire given them by the Stradivarius of flame. J.W. mused, "Man, it's hard to imagine why our lives have taken such an agreeable turn."

Branch growled, "Man, you're beginning to sound like Gillette. Would you converse in a conventional fashion…if you don't mind?"

"No, really, this is like being blessed. Ranger Branch, you know how you were talking about a state of grace? Everybody wants to know where they're headed at the end. Everyone's lookin' for a clue, yes? To me this isn't a clue; it's the answer."

"Ranger, you have lost it. If freezin' your butt off waiting for a piece of dry toast is salvation, you got a screw loose."

Bearchild smiled. "Both right, both wrong. Grace is bein' happy with what you got today, and workin' to make the clan a better place tomorrow. Now be quiet and stop wastin' calories."

The cracking of the fire was deafening, the spattering food adding to the charivari. Preoccupied with licking popping grease off their arms, they were deaf to the snapping of deadfall becoming swiftly louder. Branch leaned forward, his face glowing in contemplation as he stirred the pan with a stick and picked out chunks of egg shell. He managed two dignified revolutions before his eyes were drawn away from breakfast to the tree line. The serenity melted from his face like the lard in the pan.

J.W. turned. His first thought was to blurt that Branch and Bearchild had forced him to do it. His second was to run, but Cowsen, with folded arms, stood before the only egress from the clearing.

The first sergeant rubbed his chin quietly in deep consideration then began a gentle diatribe. "Rangers, I am truly hurt that I was not invited to your party. Do you think I enjoy partaking of fresh egg salad sandwiches, tuna sandwiches, and hoagies with an assortment of smoked meats, while you feast on such a sumptuous repast?" His voice raised several decibels. "On your feet, gentlemen. Dig!"

J.W. thought First Sergeant Cowsen was asking if they understood his command to stand up, but Bearchild slipped the army entrenching tool off his web belt and began to pick at the hardscrabble.

Cowsen marked a six-foot-by-six-foot outline in the snow then added, "As deep as Bearchild is tall, Rangers. I want a two-hundred-and-sixteen-cubic-foot hole where there is ice-hard dirt now."

J.W. and Branch pecked with their spades, spindly excuses for shovels, at the unyielding earth, netting half a coffee can of frozen dirt with each bladeful. They looked up at Cowsen, appealing with shrugged shoulders and doleful eyes, but the first sergeant's expression did not flicker. So they rasped and bitched until the hole was a foot-and-a-half deep. J.W. smiled, "Hey, this is where the earth's not frozen anymore."

Branch spit, "Yeah, well this is where the granite and the roots begin, man.

For the next two hours, as the hole assumed the shape of a mass burial pit, J.W. wondered if the scent of cooking food had addled Cowsen's reasoning, and he was going to act out what he had witnessed in World War II—prisoners digging their own graves in the POW camps. When J.W. could barely see over the rim, Cowsen ordered, "Bearchild, stand up, but not too straight. Be like Dig Forever over here. One more inch. Do it."

When the frizz on Bearchild's head was below the edge, Cowsen snapped, "Branch, Bearchild, climb out. Stand at the edge." He laughed sardonically and glared at J.W. "Ranger Weathersby here is going to bury dinner."

J.W. heard himself begging through the haze of his fatigue and hunger, "Sergeant Cowsen, sir, do you know what we went through to get this stuff?"

The sergeant spoke crisply, "Don't call me 'sir.' Serve dinner, Ranger." J.W. lifted the mess kit in which the contents had already congealed into a frozen brick of white pork belly marbled with yellow streaks of farm fresh, free-range chicken yolk, all of it waiting to be enfolded in just-baked wheat bread.

"Now!"

J.W. turned the frying pan upside-down and let his state of grace pull away. The thud was as loud as a concussion grenade.

Branch groused in a mumble, "Don't he know nothin' 'bout the starvin' children in Africa?"

Cowsen turned to Branch. "Into the hole, Doctor Schweitzer." Branch lowered himself and stared at the muddy walls.

"DO IT!" Cowsen exploded.

Branch stepped into dinner with both boots and ground the dream into humus. As he kicked angrily at the mash, the walls began to crumble, and Branch was pushed onto his knees. The dirt flowed in torrents, covering the little man to his waist. Bearchild and Weathersby stepped to the edge, but Cowsen stormed, "You two stop right there. Get out of that hole this minute, Ranger."

Branch, his face contorted in rage, exploded out of the mire on his own, dusted himself off, aside from the dirt stuck to his greasy knees, glared at Cowsen, whispered something to himself, and, as if counting to ten, suddenly relaxed. He turned back to the pit and prayed. "From dust to dust and from chicken and porker to dirt."

Cowsen scrunched his face and spat, "Quit flapping yo gums, Ranger. And what the heck you smilin' 'bout, Fill the Hole Forever? You heard me. We don't want no tired Rangers fallin' in there and eatin' that stuff—it's poison." When they had spread the flotsam and jetsam of the forest over the topped-off crater, and it was not possible to tell man had ever trod that patch of the planet, Cowsen directed them to the main platoon. "You will join your fellow Rangers who are digging in for the evening. Six-bys, in case you are attacked."

J.W. was assigned to dig with one of the men who had lost his Ranger buddy. The man asked, "What the hell's a six-by?"

"Six foot by six foot by six foot. You never dug a foxhole before?"

"No, you?"

J.W. asked, "Hey, what happened to your Ranger buddy?"

"He climbed a tree and wouldn't come down."

"What the hell was he doin' in a tree?"

"I don't know. He went crazy or somethin'. Just kept sayin' he could smell breakfast cookin' from up there. He said it was coming from the embassy, and he was going to seek political asylum."

"In a tree?"

"I don't know, man. He was nuts, I told ya. Then he started to climb up higher, you know, into the thin branches. Said he could smell bacon and eggs. He demanded to see the ambassador. That's when he fell. Broke his arm, I think. Maybe he ate one of those funny berries or somethin'."

With the word "ate", J.W.'s stomach gripped like a mad pit bull's jaws. When he jammed his hands in his pockets to wait out the spasm, he groped a cube of rubber. His brow furled. "What the hell's that?" He pulled out the lump of rescued fat and smiled.

After eight hours of excavation, fourteen foxholes were drilled into the forest. A new platoon leader was tapped with the shovel end of an entrenching tool and given a fresh mission. Second Platoon was to abandon camp immediately, but not so urgently that there wasn't time to fill the holes and cover them carefully with leaves and deadfall. The new commander stood on a stump. "When this is done, men, it will appear as if twenty-eight souls had not squandered nearly a day of their lives in a patch of remote rock no one wants."

They force-marched to attack an objective far above the timber-line. Halfway through the night, the airborne-qualified Rangers were collected in a clearing and directed to parachute into the next area of operations, a twenty-kilometer hike from their present position. The airborne troops were ecstatic, smiling, shaking hands, until word circulated that it was to be a night jump. Seven men went to Cowsen and swore the airborne list was inaccurate, that they had never been to jump school. With no way to check the list, and no way to force a man to jump who said he hadn't been to jump school, those men formed up to rejoin the legs. As they about-faced, a chorus of cadres hissed, "Pussies." A sergeant sidled up to one of the seven and asked under his breath, "You don't really want to give up the chance to save yourself a day of walking, do you, Ranger? Think about it."

The Rangers who fell into the airborne detachment left the clearing slowly, silently, relishing a reprieve from a night on their feet, though their guts roiled with the thought of a cherry night jump into mountains and forests. Each of the troops in the detachment had made their five jumps to earn airborne wings right before coming to Ranger School. All of their jumps, however, had been in broad daylight, without wind, into manicured fields larger than New York City, miles from the nearest bush or pebble. As the men walked off, several more besieged Cowsen with declarations that they had washed out of airborne school before getting their jump wings, but the first sergeant ignored them and sent the lot to the helicopter landing zone on the double-time.

During a break, the patrol leader briefed that they would be airlifted by a HUEY to a larger LZ, where they would board twin-rotor Chinooks for the actual drop. They were forewarned, "If you are late, Rangers, for the first pickup, the pilots have been ordered to presume you have been killed or captured. They will leave immediately. Then, Rangers, you will walk into the tactical problem, just like the non-airborne troops, but twice as far. You see, the airborne pick-up zone is several miles in the opposite direction of the final objective."

They pushed themselves to arrive fifteen minutes early, set up into sticks for boarding, then waited two hours until a pair of tired "B" model HUEY helicopters skidded in. J.W. recognized them as aircraft retired from combat duty in Viet Nam because of age, combat damage, or non-repairable mechanical problems. He jumped aboard the lead ship, but it was so heavily loaded, the ancient tub groaned as it struggled to lift off. The pilot pulled more and more collective, until the lever came to the end of its run, flooring the gas pedal and only skidding, going nowhere. Glued to the ground, the pilot struggled to rock the ship into a skidding takeoff, yanking back and forth on the cyclic as if in a car stuck in snow. He ignored complaints of nausea from his passengers, but when the co-pilot began to gag, he dropped the collective, pointed at J.W., and waved his thumb toward the open door. Weathersby shook his head wildly and jumped forward, making a fist around the base of the pilot's seat, refusing the aircraft commander's order to let go.

The pilot yelled over the engine howl, "Calm down, troop. We'll be back in ten minutes." He leaned over, uncurled J.W.'s fingers, then waited until J.W. had put a single leg out of the hatch to pull the collective harshly. J.W. landed on his belly in the snow.

"You *better* be back here in ten minutes!" J.W. screamed, shaking his fist.

The ships returned, but it wasn't for over an hour. When the pilot looked forward after making sure his commuters were boarded, J.W. gave him the finger. The pilot turned quickly, stared at J.W., then began laughing. He keyed the intercom switch and reached forward to the instrument panel to flick on the cabin lights. The co-pilot looked over his shoulder, his eyes dropping to J.W.'s crotch. He laughed, too.

J.W. glanced down to see the completely shredded inseam of his fatigue pants. Maybe they were laughing because J.W. wasn't wearing underwear.

The lights switched off as they gained a bit of altitude, skids slapping the treetops. The whack of the rotor and turbine whine were sounds and sensations that had been ground into J.W.'s subconscious over the hours and days and months of ass chewings as a student pilot. At that moment, J.W. felt aloof, in control for the first time in months. He knew more about this machine than any Ranger or instructor in Dahlonega, "All of you assholes put together," he snarled aloud.

He crawled up to the officer in the right seat and told him, "I used to drive helicapeters before this shit."

The aircraft commander nodded knowingly but smirked down at J.W. "Yeah? Most of the pilots I know wear clothes. Don't see no wings on your fatigues."

On final approach, one of the pilots turned and flipped a brown paper bag at the Rangers' feet, but no one dared touch it. The co-pilot had to yell at the top of his lungs over the engine howl, "That's for you clowns. Hurry up before I take it back."

Fricker was the first to pounce, slamming his fist inside like a drowning man grabbing for a log, a crazed bear. J.W. locked onto the vision of a Mars Bar materializing in Fricker's hand and involuntarily leapt forward, snatching the bag away. He retrieved a Three Musketeers then seized a Snickers and a Hershey Bar with Almonds, shoving them into what was left of his fatigue pocket. When the other men lunged for the bag, the aircraft listed to one side.

J.W. remembered neither the jump nor the landing. He was asleep for the whole exercise. On the ground, Branch stood over him and shook his head. "Man, you landed as hard as a fat lady jumping off a footlocker. You okay?"

J.W. was about to answer when they were startled by tormented shrieks. "Oh, shit, someone's hurt," J.W. blurted. "Come on, let's go."

Those who had landed safely stowed their chutes in their packs and ran toward the cries. It was Morelotto, snagged in a hemlock, legs dangling forty feet above the forest floor.

A cadre walked under the tree. He hollered up, "How the hell did you miss the DZ, Ranger? What the hell's wrong with you? We briefed you that the wind was northerly at ten knots. That means you need to steer *into* the wind, not away from it. How come you're the only one who missed the DZ? Huh? Answer that for me."

Someone piped up from outside the gaggle, "Asshole did the same thing in jump school. Shoulda kicked his ass out, but he's got pull. Now I gotta wait to eat 'cause a him. Shit."

Morelotto could not hear the conversation over his moaning. "Oh, my God, what the fuck am I going to do? Oh, my God, oh, my God."

The instructor called up, "Swing over to that branch two feet away from your face." Morelotto jiggled his legs. "No, Ranger, *swing* your legs. When you get there, next week if we're lucky, just don't get the risers all twisted up. Get goin', Ranger."

As the sergeant walked off shaking his head, Morelotto tangled himself in the spaghetti-like, nylon cords, which disturbed the balance that had supported the canopy. There was a snap of thin twigs followed by the crack of branches. Morelotto's scream sent the rubberneckers scampering from the clearing. With an abbreviated whoosh as the parachute gave way, J.W.'s mind flashed to Smith, and he waited for the thud.

Reaching the thicker, lower limbs, however, Morelotto tumbled a couple of times, becoming more deeply wrapped in the risers like a plump fly in a spider's web. The cocoon padded him as his body swung into the trunk of a tree. He swayed for a few seconds, though was unable to grab a branch, his arms so locked in the risers. One of the men reached up and undid Morelotto's harness. He worked himself free and collapsed awkwardly. He wept, "My leg's broke. Both of 'em, I think." His eyes rolled back into his head, but he managed to whimper, "I'm losing consciousness. I can't breathe." He drew in a gulp of air and cried, "MEDIC!" When there was no response, his voice faded. "Medic," then, "Medic."

With the body curled into a harmless fetal position on the ground, several Rangers tiptoed cautiously back into the clearing. Gillette warned, "Don't touch him. He could have a severed spine."

So they milled about until Sergeant Poliak pushed his way to the center of the gyrating swarm. He measured the height of the still-swinging chute and harness, spit on the ground, just missing Morelotto, and stepped around the writhing Ranger's hulk.

As Poliak left the clearing, he called over his shoulder, "What is your problem? You just completed the gentlest landing you'll ever have as a paratrooper, Ranger. As if we'll ever let you jump again." Several Rangers nodded. Then Poliak boomed, "Get your chute out of that tree and stowed, or I'll make sure you pay for it, and get your ass into formation at the same time."

J.W. studied the harness. It was hanging from a limb no more than eight feet off the ground. Morelotto had fallen about a yard.

Morelotto stood, but limped about the clearing, his face contorted in agony as he folded his canopy and tied it half-in and half-out of his pack, as had the rest of the Rangers. A freezing rain began, one that marinated the silk. Fricker raised his hand and asked Poliak where the drop-off point was for the airborne gear. The grizzled sergeant glared, but did not respond. Second Platoon carried the waterlogged chutes for another week, until the end of Mountain Phase.

After the last field problem in Dahlonega, they were marched to the huts and given fifteen minutes to pack the gear they had never unpacked. They boarded the same open canopy troop trucks that had brought them to Dahlonega.

At Fort Benning, the temperature hovered in the mid-thirties, and they were still saturated with slush from the torrent that had fallen on the downhill trip from the mountain phase. In formation, Vock and the staff sergeant held them at attention in front of the barracks until Sergeant Cowsen appeared, perfectly dressed. He ordered his charges to knock out twenty-five-and-one for being soaking wet. While they were on the gravel, he spoke. "You got leave, but you are instructed to behave like gentlemen, like the elite soldiers we are spendin' so much of our free time trainin' you up to be. I don't want no more telephone calls from Gary Rich's Steakhouse in Columbus. And when you return here in four hours, gentlemen, you will pack for the final stage of Ranger School. I would suggest you all order dessert tonight. You may have heard that a good meal is best finished with something sweet. But Rangers, do not be fooled. That may be how life ends in the restaurant world, but it is *not* how finishing Ranger School's gonna happen."

J.W. called Krista from a pay phone on main post. The line was busy, and he accompanied his platoon, shy Bearchild, to the restaurant in Columbus several miles from the gates of Fort Benning. He dialed again when they got to Gary Rich's, but she was still on the phone. He gobbled his nearly raw steak and tried again, but this time there was no answer.

At midnight, they were rousted from their bunks into formation for the final spasm, the Jungle Phase, in Florida. The drivers ordered the Rangers to deploy the thick tarpaulins over the truck bed for the trip into the tropics.

# CHAPTER 3

# MOUNTAIN GUERILLA

Ranger Jungle Training Facility
Eglin Air Force Base
The Florida Panhandle

The journey to Eglin commenced with the Rangers comfortably warm as their body heat was trapped in the canvas-cov-
ered truck beds. By the Georgia-Florida border, though, they were
pulling off sweaty fatigue shirts. Thirty miles later, some of the
men stripped off their fatigue pants.

The expedition concluded with an explosion that sent a cu-
bic yard of wet sand heaving through slashes in the tarp where
Morelotto had wielded a bayonet, desperate for air. The dirt was
followed by shards of razor-sharp palm fronds blasting through the
open tailgate from a second explosion. Though the men had been
asleep, they sprang from the floor and hard benches, banging into
each other and tripping on the duffle bags lining the floor.

From outside the cattle trucks, an unfamiliar voice shouted,
"MINE!"

A second's hesitation intervened, the remnants of 68-B pausing
to rub crust-shut eyes lest they take a sudden step and plummet off
a mountain ledge.

Fricker grumbled, "Why does the act of finishing a few minutes of sleep always have to hurt so bad?"

"MINE!"

The Brownian motion inside the trucks continued. The new voice cried, "FALL IN!"

The platoons tumbled out of the trucks, clawing their way over each other and landing at their automated stations. Gillette postulated, "I have come to the position that platoon formation is not a natural physical phenomenon. In reality, it flies in the face of the Second Law of Thermodynamics. It is counter to accepted scientific principle. We are experiencing 'The Entropic Theory in Reverse.'"

Bearchild nodded, a threadlike smile lifting for less than a second.

The company stood dripping with the detritus of their new home. In command was a Hispanic staff sergeant with a deep-colored, crescent scar that ran the length of his left cheek. He was raving about something. J.W. closed his eyes, assuring himself this was just another in the string of recurring nightmares; he was, in reality, at home in bed, cool and secure, as he had been for the past hours, the taste in his mouth a product of a dozen nights of drunken partying, not from rotting in a Soviet prison.

He looked about at his compatriots, two dozen without shirts, ten without pants. His world view narrowed as he forced himself to appraise his new overlords and speculate how they would transform a gloriously warm, tropical beach into an insult. As a few more neurons stirred, he became aware of his skin. He touched his cheek—it was moist with a thick liquid—and gathered he'd been wounded in the blast. Despite the sergeant's continuing diatribe, J.W. dropped his pack, slipped a shard of what was left of his shaving mirror from the rucksack, and raised it up to assess the damage. When his eyes accustomed to the intense light, he could make out it was just half-dried drool caked on his cheek.

The company was in formation facing the tropical sun, and though it was early morning, the glare was blinding. Fricker winced and turned his head toward the shadows. Before his neck had finished rotating, one of the new cadres bellowed, "You, skinny Ranger lookin' backwards, drop. And you with your rucksack getting dirty on the ground, you can join him." As Fricker and J.W. went down to one knee, the rest of the company began to lower themselves, but the sergeant yelled, "And the rest of you, no one told you to move."

While Fricker and J.W. were in the arms extended position, waiting for their sentence, that cadre paced circles about them, his face twisted in disgust. He broke away and wandered up and down the line of Rangers, then back to Fricker. "Ranger, if you're lookin' over there, you ain't lookin' over here. Right?"

"Yes, Sergeant."

"Now, a Ranger must pay attention in the Jungle Phase. And you with the bald head, you tired of totin' your rucksack? You want me to carry it for you?"

"No, Sergeant."

"Okay, now that we got that straight, all a ya listen up. Like I said, this here's the Jungle Phase. It's for real. The clock starts now. Fifty-and-one, and the rest of you Rangers, what are you waitin' for?"

With the entire company calling cadence, there was time for J.W. observe his new surroundings. He was struck by the Spanish moss clinging to budding deciduous trees, the thick palmetto brush, and the broad-leafed tropical growth he had seen only in movies. He glanced surreptitiously at the new cadres strutting in front of the formation, a swarthy lot, darker and smaller than their brothers in the Mountain Phase. Several Hispanic instructors stood off by themselves.

J.W. heard one grunt disdainfully, "Da reech bastards. We teach dem something."

The cadre in command, Sergeant Esposito, wore his uniform so tightly tailored, every thin streak of his musculature was flaunted on the surface of the jungle fatigues. He spit disgustedly on the sand as he purveyed his new crop of psychological fodder.

"Welcome to Florida, Rangers. This ain't college no more." He continued in a thick Spanish accent, "You may think you are almost done, but we have you for enough time to teach you something. In the Jungle Phase, you will learn to drive on."

J.W. yawned, but the man's face betrayed more than the usual disdain for trainees. J.W. noticed Esposito's Combat Infantryman's Badge, the C.I.B., a Revolutionary War musket surrounded by a wreath, and he remembered that to wear the decoration, one had to have been part of an infantry unit that'd survived real combat. J.W. wondered what Sergeant Esposito had been doing the four years when J.W. had guzzled beer at loud fraternity parties, when he'd slept through eight o'clock classes, and when he occasionally made his ROTC class in the afternoon, on Sterling College's manicured quadrangle, where they fought mock battles and snuck to the flanks to flirt with co-eds.

While most of the men on J.W.'s side of the formation would be moving into the professions, the only moving the Sergeant Espositos would be doing was of grunts, and for the rest of their twenty years. There were few occasions in the military, or elsewhere for that matter, where such a concentration of the educated and lucrative-futured found itself under the absolute control of those whom they would eventually dominate. It was only for a thin a slice of the present, but it was enough time, as the man had gloated, to make it a sobering thought.

⊨⊨ ⊨⊨

J.W. and his colleagues responded to a squawked, "Move out," by falling into their habitual tactical columns and moving to a central area where a new commander would take over. Some of the men squatted down to grab another forty-five seconds of sleep before the mass trudged forward.

Esposito pointed to a path in the brushy terrain, and they advanced without a unit leader. After a mile, at a point where they were totally disoriented, Esposito tapped Bearchild on the shoulder with a trimmed-down palm frond and handed him a sheaf of papers and a map. Bearchild looked toward Branch and J.W., lifted his hand, and motioned with his index finger toward the back of the formation. They nodded, barely moving their chins, and left for the platoon's stern. Bearchild lifted his head almost imperceptibly to Fricker, who disappeared into the palmetto on point. With a few more hand signals, the squads broke into fire teams, protecting their machine gunners and grenade launchers by placing them in the center of the flow.

When Bearchild raised his open palm, the team moved smoothly forward into the jungle. Harmony Church and Dahlonega had softened their boots, and they had to listen harder for the crunch of the Ranger in front to keep from falling behind. In a mile, they crossed out of the thick tropical vegetation onto golden brown sands from which grew emerald vines that had to be hacked away with bayonets, though Morelotto only pretended to cut brush, swinging his arms like a mad man. He had dropped his knife out of the cattle truck early that morning while slashing the canvas, desperate for ventilation. In open areas, the savanna grass was higher than the tallest man, and it took several hours to traverse a few hundred meters.

It was not simply the terrain that drew their movement to a near standstill. Myriad booby traps had been set at crossroads, and violent detonations tore at the platoon every hundred meters as they ventured onto worn paths. The cadres had deposited hunks of dynamite tied to branches throughout the Jungle School. Aggressors hidden in the lush undergrowth discharged the extraordinarily loud devices over the Rangers. Occasionally, J.W. spotted the crude, TNT-filled contrivances dangling from the limbs of the jacaranda and tung oil trees under which the platoon stopped to rest out of the sun. It was always too late, and the fake bombs showered them with atomized cordite, still-sizzling gunpowder, which stung like burning grains of pepper.

Fricker forged ahead on point, spotting and disarming several booby traps, trucking on until he came to a gentle, tropical mini-river. He paralleled it, avoiding the munitions he could see slung over the narrowest fording point. Instead, he chose a wide, but shallow, section to cross. But as he lowered himself off the bank, the water deepened abruptly. He struggled to grip at deadfall as the undertow seized his legs and carried him, tumbling, until he came to a tree branch poking half-way into the river. He grabbed it and came to a stop with an audible tearing of his bad shoulder. The pain that built over seconds was so penetrating, he drifted toward unconsciousness. In his stupor, he screamed, "Drive on!" so loud, the patrol, and even Branch and J.W. on rear guard, heard him.

Bearchild slowed and waved his palms toward the ground and crooked a finger at Gillette. The patrol fanned out into a defensive position then dropped to their stomachs, guns aimed out. Bearchild whispered to Gillette, "Go out on point. Reconnoiter, but do it slow and deliberate. I don't need to tell you that the enemy might have captured him. Maybe they're torturing him and making him yell, assuming we'll fall for it and walk into an ambush. Be careful."

As he inched forward, Gillette moved through the dense growth in a looping maneuver, approaching his stricken comrade from downstream. He returned stealthily and whispered to Bearchild, "I can see him. Just his head and shoulders. He's stuck in the mud, probably quicksand."

At the streambed, Sergeant Esposito appeared silently from the tangled vines and hissed disgustedly at Fricker, "Stop moving. Just lay there, ya dumb sh... Why the hell did you get into that crap?"

Fricker opened his eyes and tried to answer, though his voice was so faint, no one could hear him. He began flapping his good arm, then tried to scissor-kick, but that drew him more deeply into the sinkhole.

Gillette called to him, "Hey, Fricker, the density of your body is less than that of the sand. Less dense comes to the surface, yes? Like oil floats on water, right? You can float if you stop moving. Lean backward and let your legs drift up."

Instead, Fricker whacked feebly at the mire with one hand until he was enmeshed so far, only his head protruded. Esposito's face was scarlet as he took the first step to cross, though Fricker, having spent his last grain of strength, leaned back in preparation for death. The moment the thrashing ceased, he began to bob toward the surface. When his arms emerged, he stretched the working one to his side, as if half-crucified. A scorpion plunged from an overhanging puff of Spanish moss onto Fricker's hand. Though it was just a small green spider, Fricker began hyperventilating and beating his hand over his head like a cowboy lassoing a heifer.

"Leave it the hell alone," Esposito called across, but now the arm was spinning like a windmill. The sergeant raised his voice. "Sit still! It'll crawl off if you don't screw with it. And if it bites ya, you'll probably just get sick to your stomach. And while you're at it, get the hell out of that quicksand, pronto like."

Fricker managed to bring the hand of his bad arm toward the spider but missed. The creature lifted its petite tail and brought it down like a heat-seeking missile, his target a bulging vein on the back of Fricker's hand. When it got purchase, its body lifted off the skin, straight up, suspended by the tail.

Fricker screamed and slapped his arm against the sand so violently, his body lifted out of the muck, a launching Titan rocket. He sprinted over the mud onto the bank, ran in circles for a few revolutions like the headless chicken in the mountains, then stopped, vomited on his boots, and fell to the ground TU.

Several Rangers dashed across the water and lifted Fricker into a sitting position. He yelped in pain and pointed at his shoulder. One of the men undid the few buttons left on his fatigue shirt. The front of his shoulder bulged and Gillette mumbled, "Shit, it's dislocated."

They raised his right hand, which had swollen to the size of a grapefruit, above his head and called to the sergeant. Esposito forded the rivulet and stepped over Fricker without looking down.

The platoon moved across the savanna, finally entering a swamp and sloshing through stands of cypress. Fricker collapsed into the muck repeatedly. On one fall, he aspirated the scummy backwater and threw up again.

Morelotto clucked, "What the hell's he got left to urp up? Shit, the last time I ate was Christmas."

Fricker's jaw tensed. He growled, "Drive on, Ranger," and lifted himself out of the sludge. His next step was driven by bravado, the following pace a little more backward than forward; the final dropped him back into the mud. Sweat streamed from his face faster than the swamp water dripped off his fatigues. His breathing shallowed, but the rate rose precipitously as he crawled into a fetal ball. His limbs began to shake until the marsh water vibrated in resonance with the tremor. Then he stopped moving completely. J.W. was sure Fricker was gone, but his pal opened his eyes and hissed, "Drive on." His struggle to crawl out of the slime was his final volitional act of Ranger School.

J.W. lifted Fricker from the slough and lugged him for hundreds of meters, screaming at him the whole time to move his ass, but whenever J.W. let go of Fricker's fatigue jacket, his wounded compatriot lay back in the swamp water, barely able to keep his head high enough to avoid drowning. Eventually, Fricker's eyes rolled back into his head, and he dropped into a stupor. J.W. flung him over his shoulder in a fireman's carry, occasionally putting him down, but Fricker remained upright for only one or two steps. After fifteen minutes of hauling his moaning cargo through the swamp, J.W. dropped Fricker on a dry patch.

Cursing, J.W. played his last card. "Drive on, goddamn Ranger! Move your ass. You can do it if you want to," but Fricker, now unable to speak, begged with his eyes not to be left behind.

Branch and Bearchild helped carry, until they saw in the distance a clearing in the triple canopy jungle, the Eglin Air Force Base Ranger Camp at Field Seven. Fricker focused on the barracks and smiled weakly, but as they crossed the last patch of savanna, he passed into a nether world of babbling incoherence.

Second Platoon broke out of the swamp half-a-kilometer from the billets, white, two-story buildings that evoked from J.W. a memory of the first night at Harmony Church when it took eight buildings to house the company. Forty-three days later, Ranger Class 68-B fit comfortably into two double-deckers.

Dropping rucksacks at the barracks, and Fricker at the camp aide station, the platoon marched to the mess hall for a quasi-lunch of undercooked chicken and potatoes. J.W. had promised to bring something for Fricker, but when he walked to the clinic ten minutes later with greasy, pink chicken parts wrapped in a thin paper napkin, Fricker and his gear had vaporized.

"He worked so hard," J.W. argued. "Man, he was a good Ranger. There has to be some justice, some damn reward for the sweat he paid."

Branch offered his own theory. "Man, they just took him to the base doctor. Sick call. Calm down. He's too funny to die."

J.W., dissatisfied with Branch's explanation, burst from the barracks and charged back to the mess hall. He ran to a table of cadres, the Hispanic cabal. Sergeant Esposito looked up, but seeing a Ranger, returned his eyes to his plate.

"Sergeant, I want to know what you did with Fricker," J.W. challenged, a bit louder than he had intended.

Esposito looked up again, staring maniacally into J.W.'s eyes, waiting for his violent demeanor to intimidate J.W. and send him on his way. But J.W. held his ground. Esposito pursed his lips and spat with a heavy accent, "You want to join 'eeem, Ranger?"

That brought fits of sham laughter from the table, until J.W. muttered, "Assholes," but so quietly, no one responded.

"I want to know what happened to Ranger Fricker," J.W. persisted. "I carried the son of a bitch for miles. Where is he?"

Esposito rose, fist cocked by his side. J.W. dropped back a step, and at that, Esposito and his *amigos* guffawed and banged hands on the table, now bursting with feigned hilarity. One of them pulled Esposito back into his chair, and the conversation resumed in Spanish.

J.W. pointed a finger at Esposito, who jumped back to his feet and slapped J.W.'s hand away. J.W. shook his head angrily, formulating plans to return to Florida after the end of Ranger School and ambush the cabal, this time with real ordinance. It gave him the hope and strength he needed to turn and speed off to the command shack to search for Cowsen, but the duty officer had never heard of a first sergeant by that name.

At the barracks, Bearchild opined, "Fricker'll only have to repeat the Jungle Phase to get his tab. No one would make a body go through the whole thing again."

Branch was more confident. "Man, I'm tellin ya, he's just at sick call. Let him rest a few hours, get a shot of antivenom, or whatever the hell it is they give you for bug bites. He'll be back tonight. They're not gonna leave his gear around. Of course they took it to the hospital with him."

"It's antiven*in*," Gillette corrected.

In formation, Gillette pouted, "Another contradiction to the laws of physical science. No matter which way they turn us, we inevitably face into the sun. You must need a PhD in metaphysical meteorology, or some goddamn thing, to get assigned here."

Morelotto paused then grunted, "Hey, no shit."

The first dash at Eglin brought them to a set of decaying bleachers anchored in several feet of tepid swamp water. In the Jungle Phase, the balcony was reserved for the alert; only the reputed sleepers were assigned front row center. When they saw those delegated to the low seats fighting to keep themselves from floating away, Second Platoon's Second Squad sprinted smartly up to a far corner of the stands.

Several small burlap sacks were piled on a table in front of the bleachers; "Great, another fuckin' show and tell," came from deep within the peanut gallery.

A spit-shined instructor, who had found a shoal and was standing in only a foot or two of water, welcomed them to the Jungle Phase then demanded to know which of the Rangers had made the introductory comment. After several sets of waterborne push-ups, Morelotto finally raised his hand, and the cadre had him stand with his feet halfway off the back of the top rung of the bleachers, holding his M-14 to the side, the butt three inches off the seat. He wove about seeking balance but started to slip backwards, until Bearchild reached over and grabbed him.

The company was offered a welcome. "Gentlemen, you may have noticed that several men who started the course have disappeared—some even recently. Sadly, they will miss the most important training of Ranger School. All of the preparation you have been through was for this most critical of the phases. This is why there is a Ranger School. Oh, and Rangers, you are about to learn the true meaning of 'drive on'."

He nodded gravely to another cadre, who took a position at the lectern and surveyed the undulating burlap bags. He selected the one that wiggled most vigorously.

J.W.'s eyes enlarged, but Branch reassured in a whisper, "Just a sleight of hand trick, man. Whut'ch you worried 'bout? Ain't I taught you nothin'?"

"Now, Rangers, you will be meeting all sorts of creatures in the swamps. Be careful. Some of these fellas can clean your clock." The instructor opened the sack and brought forth an ugly, green serpent. The reptile flexed and extended furiously, seeking to turn its unhooked jaws toward its tormentor. The sergeant laughed. "This is a bad actor, Rangers. But don't worry, this one's venom glands and fangs have been removed. We are concerned foremost with your safety."

He charged up to the top of the bleachers and extended it to Morelotto. "Take it, Ranger, but be gentle. George here is like a Ranger. He doesn't like rough treatment."

Morelotto eyed the organism as it opened and closed its mouth, hissing puffs of wrathful air he could feel on his face. What appeared to be glistening drops of venom dripped as the head oscillated hypnotically.

"Take it, Ranger!"

Morelotto put his M-14 on the plank and reached forward with his right hand, but it was trembling, and as he grasped the triangular head, the snake slipped through his sweaty fingers. It dropped behind the bleachers into the swamp, where it squirmed in a fury under the seats to pop out in front, abeam the podium.

Sergeant Esposito, who had been standing on the other side of the stands, dove after the snake as it struggled to escape. He went one way, the snake another. The sergeant slipped, falling face first into the murk. The snake reversed course, and when it got within two inches of Esposito's face, it hooked hard left and vanished at lightning velocity into the reeds and cypress.

Morelotto, unable to ignore the blistering hue of the instructor's face, began a descent, cutting through the mass of students. As Morelotto reached the bottom rung of the bleachers, Esposito finished righting himself and dashed toward him. Eyes gawking wildly at Esposito, then up at the instructor, Morelotto whirled about, retreating back up the planks. With the instructor a pace above him and Esposito a stride below, Morelotto jammed to a stop on the algae-coated wood. Though he tried to jink left, as had the green serpent, his feet slipped, and the three collided. Esposito bounced off Morelotto, and as his foot slipped on the slimy wood, his leg dropped between the seat and the floorboard.

A crack as loud as Smith's splitting rifle butt echoed off the swamp. Esposito gasped as he looked down to see a bleeding tibia poking through skintight fatigues. The man lay silent, unmoving, and J.W. thought the sergeant was dead, until Esposito lifted himself and crawled to the second row to sit and wait for the medics.

J.W., ecstatic with the thought of a corps of cadres depleted by one, whispered in a Spanish accent as the sergeant was loaded on the litter, "Say hello to Fricker. Guess you gonna join 'eeem!'"

Esposito spit toward J.W., who dodged the hocker. The sergeant laughed.

The instructor at the podium took up where he had left off, neither adding nor subtracting a single word from his prepared lecture. He opened another burlap sack, extracting from it a beautifully-colored, orange-and-black-banded snake. Though this creature was small, thin, and docile, the instructor handled it more carefully than the pit viper. He spoke seriously. "This is the most poisonous reptile in the Americas.

"Now, Rangers, the coral doesn't have fangs. It has to gnaw its victim to death, and it has to be soft skin like in the web of the toes or hands." He held the short, vividly colored animal aloft and screamed, "You will not be allowed to handle this specimen! Except maybe for you, Ranger." The sergeant climbed into the bleachers and waved the coral snake in Morelotto's face.

The next exhibit was a repulsive, tawny viper, passed around for every Ranger to fondle. Bearchild handed it directly to Gillette, skipping Branch, who had shrieked, "Get that thing away from me. I ain't touchin' it man." It went past Morelotto, over his head, to J.W., who snatched it, pinching the serpent's head between his thumb and index finger, worried that if he dropped the handful of slimy, deadly poison, the cadres would stuff him into one of the burlap bags and hold it underwater until the end of the Jungle Phase. But the snake was cool and dry, like fine leather, and it curled around the warmth of his hand. It was pleasant, and J.W. relaxed his grasp. The snake seemed pleased and turned its head toward J.W.

"Shit, this thing's got fangs!" His hand tightened around the snake's neck so hard, it writhed wildly, sufficient reason for a prudent Ranger to fling it over the back of the bleachers, back into the swamp where it belonged.

J.W.'s arm began an upswing. The sergeant dashed along the planks and swiped the creature just as J.W. let go. "That was his goddamn tongue, Ranger, not fangs." He jammed it back into the sack and shook his head. The rest of the session was all tell—no show.

At the end of class, J.W. raised his hand. "Sergeant, do reptiles sleep at night like people are supposed to?"

"Gentlemen, night operations are the commonest time to suffer pit viper envenomation. If you encounter one of these creatures, you are to give it a wide berth."

The company split into platoons and departed on the first tactical problem of the Jungle Phase. Branch, tapped as the initial cadet platoon leader, took the manila envelope, reached into his rucksack, and pulled out rimless glasses. A black sergeant watched Branch and laughed, "You look like you know something there, Ranger."

Branch blurted back before J.W. had even digested the observation, "What do you mean? You didn't think a nig... black man be good enough to know somethin'?"

Branch turned away seething but soon took a deep breath, studied the orders, and raised his hand with an open palm. Bearchild and J.W. took the rear. Branch closed his hand until only the index finger stuck skyward then looked toward Flanagan, one of the new guys. Three strangers had shown up a few days before wearing weird fatigues and blocky caps. Unbeknownst to the Rangers, the Department of the Navy had struck a deal with the Army to send the occasional Marine Force Recon cadet or three to the Ranger School for part of their instruction. The Rangers, irritated that the Recons had not gone through the first two phases, and were there only for jungle training, gave them the cold shoulder. What most of the Rangers didn't know, however, was that the Recons had already been through a year's worth of first and second phases, and only one out of ten of their starting numbers had made it this far.

When Branch's finger rose, he expected Flanagan to stand, mouth agape, and trigger bellyfuls of arrogant laughter from the real Rangers. The new man, however, blinked his eyes in acknowledgement and slipped out of the clearing onto point.

The triple canopy jungle covered a stinking marsh dotted with the rounded stumps of young cypress trees. Further into the mire stood the tall, thin trunks of mature cypress, their dark, feathery leaves forming a gloomy roof over the swamp. What meager light filtered through was absorbed by the Spanish moss that hung so thickly, there were times, even at noon, the troop to the front was only an eerie shadow. The jade, velvety web of moss, though beautiful from a distance, entangled easily in clothing and weapons. A Ranger's open mouth activated a cosmic attraction between the intensely bitter moss and his tongue. Once in a Ranger's mouth, the next hour was given to picking the fine, hooked tendrils from mucous membranes.

The swamp water was a different matter. Though the cadres warned that if even a drop of the slough liquid was swallowed, the ensuing illness would not be survivable, a mouth rinse of swamp water was the only way to loosen the tenacious, caustic Spanish moss. But the bog water also left an unsavory, lingering aftertaste that made not brushing one's teeth for a week agreeable.

In the swamps, one had to follow the splash and gurgle of the footsteps to the front, for the territory was devoid of snapping twigs. Rangers who fell asleep on the move soon tripped over cypress stumps to wake with gulps of putrid water leaking down their throats. That was followed immediately by fits of brutal coughing typically seen only on TB wards.

After their submerged feet had become sufficiently wrinkled and white, Sergeant Perez, Esposito's replacement, stopped Branch and suggested they leave the swamp and cross back into the savanna. He tempted Branch, pointing to the thick plains covered with endless, thirteen-foot stalks of beefy elephant grass.

"Dry up there, Ranger. And see those trees in the fields, the skinny ones? You can send a man up one of 'em and look out *mucho* sees."

Branch listened politely then gathered the platoon into a huddle and drew on his hand with a pen. Intelligence indicated the aggressors were occupying a swamp-bound compound. His plan to overwhelm them was simple: launch an attack two hours earlier than expected by keeping his men moving through the fetid swamp instead of the longer route via the bone-dry, and enticingly more pleasant, savanna.

Branch solidified his final battle plans, raised his fist, and was about to call out, "On me," but was stopped in his tracks by Sergeant Perez, who took a palm frond, slapped him in the head, and declared Ranger Branch killed in action by a tarantula as big as a dinner plate. Branch gulped. "They got them things here, too?"

The sergeant nodded. "You bet. Kill one or two Rangers every class."

Gillette laughed. "Actually not. The family Theraphosidae are essentially non-lethal to humans. No record of one killing a man. You're still quite alive."

As Perez took a fuming step toward Gillette, he noticed Morelotto asleep in the crook of a cypress. He detoured to the tree and slammed the broad leafed palm on Morelotto's head. Plans changed dramatically. Morelotto ordered the platoon to move across the dry plains and avoid the swamp at all costs. Perez smiled. Morelotto surveyed the land. From a distance, the savanna was just grass, an arid, unkempt lawn.

Though initially deliriously happy with the absence of water, the broad, sharp blades of elephant grass soon sliced skin with knife-like precision. After an hour tramping the untamed, fiendish grass of Florida's panhandle, Second Platoon was forced to draw hands up into the sleeves of their fatigue jackets and scrunch fingers tightly into fists. Fingertips became so numb, several of the sleep-walking commandos had no idea they had dropped their M-14s.

J.W.'s fingers were as deadened as if he had slept on them for weeks, and he, too, dropped his M-14, but in his case, the metal barrel struck a rock and bounced up against his right shin. He regained sufficient consciousness to see blood seeping through the remnants of his fatigue pants. He did not, however, spend the calories to bend down and inspect the wound. After an hour, with the blood coagulated and his pants glued to the laceration, every step pulled painfully on the crusted gash. When he could no longer stand it, he took a deep breath, closed his eyes, and yanked the pants free. But this only restarted the bleeding, and an hour later, his pants gummed to the laceration again. On the third cycle of pain and pants pulling, he cut a rectangular piece of plant flesh from the eight-inch-wide blades of savanna grass and wrapped it around his leg. The humble dressing kept his pants away from the wound, and also allowed the blood to run freely down the tibia into his boot.

Those who emerged from the jungle without their rifles were sent to appear before a panel of cadres, Sergeant Perez at the center. He challenged each man: "Ranger, you been carrying that piece for two months. What the hell do you mean you don't know when you lost it? Now you and your platoon get your asses back into that terrain and find it ASAP. I don't give a shit if it takes a week. I ain't in no hurry. You don't finish the course on time, we'll keep you here until you do."

By 2300 hours, with all weapons back in insensate hands, Morelotto steered his charges back into the swamp. J.W. thought incessantly of the green snake and its viper siblings, more concerned with the venomous fauna than his exhaustion. He was not alone. Each time the platoon stopped, those not involved in planning the tactical problem climbed into trees to escape the water. In the moonlight, J.W. saw Rangers propped high in the cypress, resting like night jungle creatures ready to strike. However, even the fear of death by serpent was insufficient to keep them awake, and one by one, bodies dropped into the swamp, gurgling angrily, "Fuck me, fuck this shit. I hate this fuckin' place."

Deeper into the muck, Branch, Bearchild, and J.W. were assigned rear patrol. They moved as lone foxes. At 0200, they passed a black figure, surely a bear, and J.W. whispered to Bearchild, "I think it's your cousin, or somethin', man."

In fact, it was Morelotto crouched in the slime, babbling to no one in particular. Somehow, he had managed to stay plump, still a Pillsbury Dough Boy with a greasy stubble of black hair. Branch shook the man awake. Morelotto jumped up and kicked at a cypress stump then cursed, "Goddamn machine!" He turned heatedly to Branch. "Hey, man, gimme a nickel. You owe me the money. You're the one who broke into my footlocker last night."

"We were on the cattle trucks last night, dickhead," Branch snapped, though with a sly smile

"You black mothers broke into my footlocker."

Branch's fist cocked, but Bearchild grabbed it then locked Branch's eyes. "Give 'em the money."

"Fuck, no,"

"Then I'll give it to him," Bearchild groaned as he picked a pebble out of the water and handed it to Morelotto, who snapped it away. He tried to shove the stone into the stump, though it remained recalcitrant. Morelotto punched the wood with both fists, tried the pebble again, then turned back and yelled at Branch, "I asked for a quarter. You gave me a goddamn nickel."

"Don't blame me, man. Bearchild gave you the money."

Morelotto targeted Bearchild. "The goddamn Coke machine's twenty-five cents. And you know it, Chief."

Branch corrected him. "Don't be an asshole, man. The machine's stuck. That's the problem. You need to kick it a few times as hard as you can. Hit it with your hands. Punch the shit outta the mutha fuckin' sum bitch. And don't call Ranger Bearchild 'Chief'."

Morelotto swung his rifle at Branch's head, but Bearchild stopped the weapon with his hand, and J.W. pulled Branch away.

Bearchild whispered evenly, "So, Morelotto, who's minding the store? You're the spittoon leader, right? You gotta get back up there and lead us into battle. We need you, man."

As Bearchild tugged him gently toward the main contingent, Morelotto emerged from his psychosis and walked peacefully forward to retake command. Morelotto marched at the head of his unit for an hour, much of the time spent snarling orders, until he ran out of breath and dropped back. He dragged along just fast enough to stay a few meters ahead of the rear guard, though Branch sprinted ahead every few minutes. "Think of it, Morelotto, a large pizza, man—black olives, Italian sausage, pepperoni." He'd let a minute go by then add a topping or two. "Onions, mushrooms," he smacked his lips, "anchovies, extra cheese." When Morelotto started to weep, Branch played his trump card. "Cold beer, Morelotto—five, six, no, seven glasses of cold Schaefer beer to go with the pizza. Much as you want, man."

J.W. believed the last molecule of Morelotto's biological energy had been tapped, and it was time for him to merge into the ranks of the disappeared. Morelotto dropped to a knee, and the three passed by. J.W. wanted to drag him along, but Branch whispered, "Let him be. The cards'll fall where they will."

The three men on rear point stopped frequently to rest, speaking of Morelotto and how racism had taken its tariff. They also bragged that they had evolved into nomads of the swamp, men capable of going without, toughs who existed on a plane far above that of the fodder in the ranks. J.W. set his jaw. He acknowledged, for the first time in his life, that maybe he was better than at least one person he knew, and as the three moved on, he sloshed with a swagger that lifted his spirits. He became so wrapped in his new persona he did not hear the slow, deliberate, near-silent steps from behind. In minutes they became aware of the thick splashing of a full patrol closing on them from the rear.

Branch cursed, "Shit, I knew it. We shoulda hooked south into the savannah."

J.W. nodded in agreement, as did Bearchild, the three cursing that in belittling Morelotto, they had failed to swing back around through the savanna to ensure no one was tailing them. When the flopping came closer, they dropped to their knees, preparing to jump the intruders.

Bearchild whispered, "Too many of 'em. Must be a squad to make that much noise. Don't move. Let 'em go by, and we'll hit 'em from behind."

But only a single body crashed through the swamp past them. Branch laughed derisively. "Green olives, Genoa salami."

Morelotto barely turned to flip them the finger as he jolted through the water, stomping back to his position at the head of the platoon. He reported to Sergeant Perez, "No enemy troops back there."

"How do you know, Ranger?"

"I checked myself, Sergeant." He started to add, "Sometimes you can't trust those..." He swallowed the remainder of his declaration.

"Okay, Ranger, tell your men to take a load off while I go back and bring in the rear point."

As Perez moved toward the patrol's rump, he found the three tail gunners sitting against a tree smoking, laughing about how close they were to riding themselves of Morelotto. He popped from behind a Yucca bush and had them into the swamp for fifty-plus-one, though with a jungle twist in which the long stroke was not the painful arm-extended phase, but in the low position—face in the putrid water.

Perez explained, "That's for not paying attention to someone sneaking up on you. And Rangers, that'll also help you learn how to hold your breath for long periods. A Ranger never knows when that'll come in handy. Now get your butts back to the main platoon."

Branch snuck up to Morelotto and whispered from behind a tree, "Double thick crust, prosciutto."

Perez overheard but did not rebuke Branch for the misdemeanor of Ranger goading Ranger, only for the felony of ranting about food. That cost Ranger Branch and his fellows, excluding Morelotto, a march back a few hundred meters into the drink for another fifty-and-one.

⇒╬ ╬⇐

The new platoon leader's objective was to attack a guerrilla command and control center guarded by an elite enemy force. Intelligence indicated thirty to fifty well dug-in, heavily armed troops with automatic weapons. That, however, was the extent of J.W.'s recollection of the briefing. The rest was lost in the steaming scent of the hot can of Beef Stew being consumed greedily by Staff Sergeant Perez.

Earlier in his army experience, J.W. had found the stew to be a particularly nauseating C-ration selection, on a par with his Aunt Gertrude's Hungarian goulash, or Branch's fantasy, Ham and Limas. Now just the aroma of Beef Stew precipitated one of his more and more frequent schizophrenic breaks. Looking about the clearing, he was sure everyone, not just Perez, was choking down

Cs. He stumbled toward his compatriots, though as he challenged them, the meals vanished, leaving only fictional scents of Franks and Beans, Ham and Eggs, and Spaghetti and Meatballs. Some of the troops looked up at him in their near-sleep. "What's your problem, man?"

When he tried to answer, all he could whimper was, "I just want my share of the chow. Stop tryin' to hide it from me."

As his hunger deepened, so did his despondency. He became convinced that he had reached a permanent state of painful starvation from which he would never escape. He began to surrender to the biting emptiness in his gut and his heart, and the old thoughts of quitting roiled up from his subconscious.

He announced to Branch and Bearchild at the next checkpoint, "I'm gonna sit this one out. You go on. I'll catch up."

Branch shook his head irately. "Sit this one out, my ass. Drive on, dud. Plenty time to rest when you're dead." Though the words did not motivate J.W., Branch's slap in the back of the head did, and J.W. rose to his feet to weave behind his partners. In a few minutes, though, he drifted back into semi-consciousness, falling victim to a profound food hallucination. The only difference between Morelotto and J.W. was ethnic memory. In front of J.W.'s eyes were not pizza and beer, but a plate of hot, buttered egg noodles dusted with crystals of sea salt and a few nuggets of freshly-ground pepper. In the center were two plump, kosher hot dogs flown in from the grill at Nathan's Deli on Coney Island.

J.W. drifted alongside Branch and asked if he knew where he could find Goulden's mustard to complete the feast, but Branch was busy chattering to himself, and in the haze of his nightmare, J.W. thought he heard Branch praying for ribs and cornbread. Bearchild also maundered to himself, but he named foodstuffs the ilk of which J.W. had never heard.

Weathersby chased his apparition for hours, the steaming pasta and glistening franks emanating a luscious fragrance that drew him incessantly forward. Several times his arms reached out for the meal, but his fingers were always several inches short of seizing it. The vision, so close yet so far, had become painful, and J.W. thought only of quitting Ranger School and calling his Aunt Gerty to beg for a bowl of her goulash.

Eventually the visage of his family faded, and J.W. chased frankfurters and noodles until his feet slipped from under him, and he landed on his back in a yucca patch. One of the toothpick-sized, spear-tipped thorns broke off in his butt, shattering the prospect of ever overtaking the mirage.

As Branch pulled him to his feet, J.W. begged, "You gotta help me, man. Something's stuck in my behind. Really."

"Man, what is it with you? Shit, pull down your pants."

Puzzled as to where his rear guard had again vaporized, Sergeant Perez popped into their clearing just as Branch extracted the remaining fragment of pricker from J.W.'s butt. Branch's operative efforts were recorded in Perez' notebook, who groaned disgustedly, "Pull up your pants, sick Ranger."

Back at the main contingent, J.W. watched, his mouth watering, as Morelotto, now the forward scout, reported a small group of aggressors sitting around a campfire two hundred meters ahead.

"They're surrounded by boxes of C-rations, cans of Coke, Mr. Pibb, Orange Crush, 7-Up, and a small carton of Hershey Bars, some with and some without almonds. The box appears to be half-full, and the cartons of Cs have hardly been touched."

The platoon leader wrote notes and nodded his head in thanks for the valuable intelligence, but Sergeant Perez interrupted. "What about their weapons, Ranger? What about security?"

Morelotto thought for a moment then apologized. "Sorry, Sergeant, I guess I forgot to check on that."

The enemy pretended not to hear Second Platoon approach, the Ranger lead elements sliding relatively silently through the water, drawn to their objective by an enormous bonfire.

"Man, these guys are stupid," Morelotto shook his head. "A fire like that? Shit, you can't miss 'em."

"Man, it's you who's stupid." Branch muttered. "Bigger the fire, sooner we find 'em, sooner the attack's over, sooner they get to sleep, sooner we got another damn mission. They win; we lose."

The aggressors were silhouetted by the fire, lounging comfortably on the wicker porch of a tropical house, suspended by stilts, two feet above the swamp. The Ranger platoon leader raised his hand. Safeties clicked off, and as his arm dropped, Second Platoon commenced firing. The aggressors vaulted out of their hammocks, startled, feigning confusion and disorientation. The attack was so fierce, there was no escape for the enemy. Each of them died in place, melodramatically, until the last man crumpled onto the dry porch. Their lieutenant appeared, had the deceased stand, counted noses, and told them to retreat to their rallying point in the swamp.

Most of the Rangers, believing they had crushed their adversary, dropped onto the abandoned hammocks to rest and savor triumph until the cadres appeared out of the night. "Rangers, pretend these are North Vietnamese regulars and Viet Cong guerrillas. You need to hunt them down. They're dead tired and hungry. They don't even have helmets. They're bareheaded, Rangers. That tells you they're running for their lives. Their discipline is gone. All that training, the months and years. All of it gone. Don't let up on 'em, Rangers, pursue and destroy!"

Those Rangers in the visual purview of the cadres, that is, those not hiding under the building, reentered the swamp, slogging through the swamp until a lane grader called them together half-a-mile beyond the objective. "Okay, that's it men. You did a good job. In Viet Nam, after an attack like that, you will come to cowering enemy troops begging to live. Some will run from you and trip over their dying compatriots. You'll come upon others weeping like babies, wondering if they're going to live to see the sun.

"Your job as soldiers is to fight, not murder. You are still human beings, superior human beings, and you will act as such. The enemy, once beaten, is not to be tortured or spit upon. To kill a captured soldier is murder, murder, not combat. And you are not murderers. That is not who you are, slaughterers. You are the chosen. You will pass the knowledge we are giving you on to the others you command.

"When the battle is over, you will treat your enemy with respect, for they are soldiers, no happier to be there than you are. Don't any of you ever forget that."

Those Rangers who had hidden from the cadres at the original objective forced their way into the jungle house, where they broke into shrieks that reverberated through the swamp. That drew most of the Rangers away from the sermon, back through the swamp toward the building where the platoon leader watched from the door, eyes darting. As the cadres materialized from the mist, he bellowed at his men to abandon the house and form a perimeter. The more vehemently he hollered, the faster the returning Rangers galloped up onto the porch, through the door, and into the one room.

J.W. watched as a corpus of Rangers gelled in a loop, tugging at an object at the center. He gasped as he realized they had captured an aggressor and were gutting him. He was repulsed, and despite his hunger and fatigue, dove forward into the fracas to stop the outrage. When he clawed his way into the middle, he found his fellows tearing, not at an enemy troop's remains, but at cases of C-rations, pulling the cardboard apart with their teeth, handing the spoils of war over shoulders to greedy hands.

J.W. reached for his share, seizing an obnoxious little can of bad tasting, under and over cooked, lumpy, incorrectly spiced, not spiced, unpopular selection of military "food," Ham and Eggs, one of his favorites. The more reasonable troops satisfied themselves with a single can and sprinted into the swamp to suck down the calories. A few pawed for an extra tin, their egress delayed just long enough to allow the return of the deceased aggressors, who squeezed into the hut and pulled the Rangers violently off the cache. Most of the avaricious Rangers wound up with nothing, though a few held desperately to a can as they were heaved through the door by an enraged enemy. Most of those Ranger bodies landed, face first, in the swamp.

J.W. absconded with his tin clutched in a death grip as he hurtled the railing into the slough. With feet firmly mired in the mud, he half-squatted and placed his P-38 can opener on the lid, greedily anticipating the first morsel. As the metal was pierced, however, the vacuum was so strong, it blew the little can opener out of his swollen fingers. He felt around in the water, but it was futile, so he pried open the rest of the top with his bayonet and then his teeth. The process was over in seconds, including licking the inside of the can. He leaned back against a cypress stump and burped for the first time in weeks then relived the entire encounter, adding and subtracting, dividing and multiplying, until he settled on a figure of three-hundred-and-seventy-four calories.

Some of the Rangers hid by squatting under the water, holding cans above their heads. The cadres sloshed through the swamp, slapping and kicking C-rations out of their hands. The company was ordered into formation, where the troops braced in three feet of water.

A cadre screamed, "You will open your mouths for inspection. Those of you with food on your scummy teeth will have their name recorded in this here little military notebook."

J.W. beat the cadres. He had rinsed his mouth with swamp water, gargling out rancid rodent and bird parts along with bits of illicit C-ration, laughing at the stupid cadres—it wasn't necessary to inspect mouths to determine who had eaten. The fortunate stood less slumped, their muscles not shuddering in vicious tremors.

J.W.'s body absorbed the calories abruptly, and he felt wonderful, more alert than he had in weeks. Within five minutes, though, his nervous system sucked from the pool of digested fuel sufficient energy to function, and he became more attuned to the pain in his legs, the burning lacerations on his arms and, mostly, to his biting hunger. In ten minutes, J.W. Weathersby was again ravenous, and this time he was sufficiently alert to realize it.

———※———

The new platoon leader, Wardally, an exchange lieutenant from the Grenadian Army, dispatched Bearchild, Branch, and J.W. to forward point, asking them to scout ahead and rendezvous with him several hours later. The three ventured forth on their sortie, and J.W., with what he imagined as renewed strength and a clear mind, abruptly laughed. Branch and Bearchild stopped and eyed him suspiciously.

Branch snorted, "What's wrong witch you now, man?"

Weathersby laughed again. "Think about it. This is a dog and pony show we got goin'—three loose cannons on auto pilot, amusing ourselves. We are board-certified point men, rear men, legends in our own time. Unbelievable. Even the foreign students look to us for leadership."

"Leadership, my ass," Branch taunted. "Man, I gotta get outta this water; it's burning my thing."

J.W. offered back, "What the hell you talkin' about? The water's only a foot deep."

"Well, *my* thing's draggin' in it. Mine's more than a quarter of an inch. No wonder you had so much damn trouble with the ladies."

Branch let out a howl of laughter, and even the corners of Bearchild's lips curled.

"Yeah, Bearchild," Branch cackled, "you know what I'm sayin'. Looks like one of your braves scalped him in the wrong place. Musta took it home to his squaw. Man, I can hear her laughing now. Put that thing on her necklace along with the other peas and beans and corn kernels. Plant it in the spring. Who knows what the hell you gonna get? 'Husband, you see this little plant a comin' up outta the ground? What the hell is it? A infant carrot or somethin'?'"

Branch went on and on until an event occurred in Ranger School Class 68-B that had never before been witnessed—Bearchild chuckled.

The water was aggravating, and feet were tender from the unending hours in the soup. The three consulted their maps and saw that the only dry ground within ten square miles was a dirt road on a causeway that had been crafted decades before by the local Civilian Conservation Corps. If they slogged the swamp, as they had been tasked, it would cost them five hours, but only three if they climbed onto the dry land of the causeway.

J.W. warned, "That road's marked CCC. May not still be there. Depression was forty years ago. Those guys didn't exactly build things to last."

Bearchild went back to the map. "You're right. Might be a lot of wasted time."

J.W. shook his head. "Let me take another look." J.W. measured the distance with his pencil then went to the legend. "Yeah, might be, but it's only an hour there and back. That's all we lose, one hour. Probably something still there to walk on." He made some more measurements, scribbled numbers in his Ranger Manual next to the recipe for chicken stew, and looked up. "Worst case scenario, gents, one-hour sleep, best three, and that's adding in doubling back at the end for a briefing. Worth the effort. We win every way you slice it."

Branch stared at J.W. "I thought you didn't know how to read a map."

"I don't, but I come from a family of accountants. My sister's a bean counter for Columbia Pictures, and now I'm countin' beans in the jungle. For her, beans are jewelry and money, in that order. For me, it's calories and sleep, in that order. But countin's in our blood."

They came to the dry, raised road in thirty minutes and sat there, boots off, letting their feet breathe in the night air. They changed into dirty, but dry, socks, then let their heads droop. When they heard the platoon bashing the slough, they paralleled the main body, moving slowly through the pitch blackness, chatting and smoking. A mile later, they happened upon several dark, truck-sized storage containers.

"A curious find, to be sure," Branch mused as he moved cautiously toward the metallic, oily-scented silhouettes. He gasped, "Ha!" then broke ranks to disappear into the night. Bearchild and J.W. stopped dead, standing motionlessly until Branch reappeared and whispered loudly, "They're Air Force trucks, man. Deuce-and-a-halves. Come look."

Branch vaulted aboard the lead vehicle and checked to see if the drivers were asleep. "Empty!" he called out, waving his comrades closer.

J.W. climbed into the cab and rifled for food under the seats and in the engine compartment, though he hit upon only three packs of Winstons. "My brand, ain't that somethin'!" he whispered to Branch, who shot his hand forward.

"Well, they just became *my* brand, too. Give 'em here."

In the starlight emerging from behind storm clouds, J.W. watched Branch eyeing, then playing with, the ignition switch, flipping it back and forth gently between his fingers. J.W. smelled the wheels of the man's brain grinding, his black eyes beginning to shine.

Branch whispered out the window, "Bearchild, get your butt up here."

As his Ranger buddy hopped into the cargo bed, Branch brought the engine to life, nursing the truck out of the camouflage.

J.W. stammered, "You sure this is a good idea, man? You ever heard of actions begetting consequences?"

"There you go again. Acting like a man ain't got no balls. What are they gonna do to you, Ranger? Send you to Viet Nam? Maybe they gonna send your ass to Ranger School—God forbid. I thought we already went over this. You were going to change, remember?"

J.W. did not answer, and with lights off, they crawled along the swamp-lined causeway a mile or two per hour. When one of the tires slid into the swamp, Branch cursed and jammed the vehicle into four-wheel drive, but unable to distinguish the edge of the road from the swamp, floundered further into the muck. The truck tipped sideways, the front wheels sinking deeper into the mud.

J.W. and Bearchild were about to abandon ship, but Branch called out, "Screw it," flipped on the lights, caught sight of the edge of the road, and jammed the truck into low-low four-wheel drive. He reversed, pulling out of the water, then slammed the gear shift into first, gunning the tachometer to red line. Back on the causeway, with a few grunts and pumps of his muscular arm, Branch managed to ram the transmission into cruise mode, and they sailed, pushing thirty miles an hour.

The wind was so hard in Bearchild's face, he turned away. That was when he spotted headlamps behind them, pulling closer. He leaned into the cab and warned Branch, who shifted into high then floored the rattling diesel, barely squeaking out another kilometer before skidding off the road, the front wheels miring again in the swamp. He brought the truck to rest sideways, preventing a tow vehicle from gaining purchase to pull it free, then killed the engine but left the ignition switch, the brights, and the heater motor on to sap the battery.

They jumped into the muck and set out across the swamp toward the objective. Branch sucked air like a racehorse but laughed in a howl, the first and only time J.W. ever saw him happy. Then Bearchild smiled, and J.W. waited for another guffaw, but causeway high jinx did not rise to the level of J.W.'s challenged reproductive apparatus. They stopped in a patch of tamarind and hid amongst the delicate, lacy leaves. A few shriveled, brown seedpods hung tenaciously from the branches, having survived the winter. J.W. wondered why those few had managed, and others had not.

They smoked the Winstons, watching from a distance as the second deuce-and-a-half arrived. Men stood on the elevated road milling about as if cops at a crime scene—others peered into the night swamp. One man pointed to a different sector of the terrain and took a step toward the edge of the road. He raised his arm and shook a fist at nothing.

Branch's eyes widened, as did his grin, when the aggressor crew attempted to stretch a too-short jumper cable toward the stalled deuce-and-a-half. "I should be on the Ranger School Planning Committee."

The three could hear the drivers cursing as the airmen slogged into the marsh to remove the battery from the dead deuce-and-a-half. Branch rubbed his hands, his face deadly serious. "Dumb Air Farce fucks. No way for 'em to know, man. No way. They'll never figure who it was that did 'em. Cocksuckers."

The three plodded across wide stretches of swamp in the general direction of the objective, the kilometers filled with grumbles about the stench of the putrescent, viscous water, the miles left to suffer, and the airmen who had curtailed their motorized junket.

Branch moaned, "We shoulda pushed on for another mile or two before ditching the truck."

J.W. was about to agree when Bearchild put his finger to his lips then cupped his hands behind his ears. An eerie howl skipped along the surface of the slough.

J.W. asked, "Bearchild, what is that, my man?"

"Voices."

Branch added, "They're callin' us by name, tellin' us to rejoin the platoon." He thought for a moment. "No, man, I'm not doin' it. We don't go back. We'll just tell 'em we never heard a thing." Bearchild and J.W. nodded in agreement. "They can't prove nothin'. Just drive on, man. We're doin' our job, right? Lookin' for the bad guys and keepin' them away from the platoon."

Deeper in the swamp, they came to a thatch of twisted cypress, in which they hung for an hour's nap. When they could hear the rest of the platoon nearing, they shinnied down and drifted toward the sound of men lapping through water. They huddled and agreed it was time to check in. It took half-an-hour to reach the platoon, now just twenty-eight strong, circled in a wet perimeter. The three nodded in unison as they delivered a report that they had spotted the enemy in the swamp, but the aggressors weren't anywhere near the platoon, actually moving away from the Rangers, so they didn't think it necessary to come back and warn them.

But even in the dark, it was hard for the point men to ignore the angry stares of their compatriots. At first, J.W. believed it was because they had rested, relaxed faces, though he soon learned that Platoon Leader Wardally had been relieved of command right after he had been pressed into doing a roll call in the middle of the swamp. When he was unable to account for every single man, he had been directed by the lane grader to dispatch a patrol and retrieve the point, even if that meant using the road and flashlights, or screaming names for the rest of the night. When the rescue patrol became lost, Wardally sent out two more men, who disappeared. He confessed he did not know what to do, an admission that cost him his job, likely his place in Ranger School, and perhaps his future in the Grenadian military.

J.W. argued with Branch, "Man, this is bad. What did Wardally ever do to you?"

"Nothin'." Branch shook his head in disgust. "That's not the point, man. They're not going to toss the guy just because of one failed patrol. Lotta guys fail patrols. Then they come back, do good, and all's forgiven. And, it's not gonna help Wardally for us to fess up. They can't prove anything. Nothin'. Coulda been another platoon, maybe some locals tryin' to steal it. Like I said, we just tell 'em we heard 'em callin', yeah we heard 'em, but we thought it was the aggressors, and we were too disciplined to compromise our position. Be cool. You white dudes worry about everything. Just keep your mouth closed, 'Mister Talk Forever'."

J.W. asked, "Bearchild, what do you think, man?" Bearchild looked at Branch and nodded in agreement.

A fresh lane grader appeared out of the savannah toting the five lost men. He pushed the reconstituted platoon toward the enemy position, and at dawn, he called a formation and welcomed them to the first full day of the Jungle Phase. J.W. stared at the fiery morning sun and dug into the labyrinthine interstices of his mind to imagine anything in life as woeful as starting a new phase.

⟞⟝ ⟞⟞

J.W. walked the rest of the morning wrapped in melancholy. He was locked into a closed universe in which every move he made to lessen his own pain rippled through the lives of twenty-some other men. He spent erg after erg of psychic energy atoning, in his mind, for the damage he and his two pals had wrought the night before, but after a few personal castigations, the guilt fizzled, and all he could concentrate on was his vacuous stomach.

The new platoon leader moved them through the savanna. As the sun arched higher and hotter, the troops were ordered back into the swamp. By mid-morning, they had beaten through only four kilometers, less than two miles. There were still twenty-one klics left to the next objective. The major impediment to movement was no longer the fetid water but the constant, harassing fire from aggressors who sniped at the platoon from behind, slowing the Rangers to a snail's pace. Despite the hours of scorching, humid, fake combat, by noon they were no closer to breakfast than they had been the night before.

J.W., Branch, and Bearchild grumbled unceasingly at the manner in which the platoon moved by fits and starts. It was the first time in weeks the three were not on their own, having been demoted to travel with the main body of the platoon. The order

269

came from Lieutenant Colonel Bloom, commanding officer of the Ranger battalion, who had been woken by his CO in response to a compliant from the Department of the Army, who had been alerted by the Department of the Air Force, who had been informed by the commanding general of Eglin Air Force Base, that there had been the theft and destruction of several Air Force vehicles and aircraft on the patch of wasteland generously donated by the Air Force to the Army for Ranger training. "Somebody's ass is gonna fry."

Branch moaned, "Man, our days as God's chosen are over. Prepare to suffer."

At noon, with the pace of the attack slowed to less than a kilometer every two hours, a new platoon leader was baptized. His first act was to summon the three poker-faced, restrained, former luminaries to the front of the column. He assigned them rear patrol, though the lane grader looked askance and asked, "Do you really want to do that, Ranger? A leader's supposed to know who he can trust and who he can't."

"Sir, they're the only ones who can stay awake. At least you can trust 'em not to get lost."

Waiting until the platoon was well into the swamps, the three quickly resumed their accustomed modus operandi, spreading out to grub for plants or berries, something to swallow. J.W. followed a worn path to a lone tree whose drooping stalks held brown pods. The leaves were early-spring green, some long and thin, others heart-shaped.

"Fresh produce," he called, "I'm gonna make me some fruit compote," then climbed twenty feet to a branch near the top, harvesting as many of the three-inch pods as he could stuff into his shirt. Harboring visions of roasting them over a Bearchild inferno, he summoned his friends with the Ranger Bird and proudly displayed his find. Bearchild shook his head, smacked the harvest out of J.W.'s hand, then squashed them into the ground.

"Tung oil beans. Kill ya—fool."

J.W. found nothing else for hours, though most of his day was spent washing and rewashing his hands. At two in the afternoon, he spotted an ill-fed bullfrog sunning itself, legs splayed across a cypress stump. It heaved J.W. into the latest iteration of food fantasy, one far down the menu from Nathan's hot dogs. He saw visions of being eleven again, his family's trip to Montreal, and the dinner in an up-scale hotel. His sister ordered a sandwich, *fromage grille*, melted cheese. J.W. opted for the *cuisses degrenouilles*, simply because they cost four times what his sister chose. The waiter smiled unctuously but became quite serious when J.W.'s father demanded, "What are green wheels?"

"It is ze foot of ze frog!" the waiter clicked his heels and bowed. "*Tres delicieux*, ze most delicious."

"J.W., you will finish every last bite. Do you understand?"

The grilled cheese came first. Minutes later, the waiter dropped a covered plate in front of J.W. and, with an arrogant snap, lifted the silver cap. Skinny little femurs, tibiae, and fibulae—shiny, slithery bits and pieces—elicited in J.W. a series of dry gags. His stomach was empty—his sister's was not.

J.W. waved to Branch then whispered, "Lunch is served, my man. This is gonna be good. I know about this shit."

Branch tracked quietly to the frog's flank. With a hand signal, he sent J.W. to the front as a diversion. Branch crept within four feet of the amphibian hors d'oeuvre but tripped on a hidden underwater log. The frog was airborne in a nanosecond. It ducked under a scrap of deadfall and disappeared beneath the algae. When it surfaced twenty meters downstream, Branch charged through the slough, a maddened hunter. The bullfrog sat immobile until Branch was five feet away then darted under another log, popped up a dozen meters down-slough, croaked once, and was gone.

Branch cursed the frog, and then his life, but he soon leaned back against a log next to Bearchild and fell into a trance, floating half-suspended in the putrescent water. J.W. dropped next to them and placed his hand on Branch's upper arm. He reassured, "Don't worry, man, we'll eat soon."

Branch was the first to wake. He shook his two companions, whispering he had spied a bird sleeping beside a log. Both Bearchild and J.W. followed him to the reposing dove, but Bearchild turned away, his face twisting in disgust.

J.W. shrugged, "I don't know. Maybe the bird's on his totem pole or somethin'. As far as I'm concerned, man, that's breakfast and lunch, and I'll be happy to wring its neck, build the fire, and cook the sucker."

Branch followed J.W., employing the same hunting tactics that had nearly bagged the frog. Weathersby moved with perfect silence, tasting the roasted magnificence of waterfowl, though he admitted to himself he'd happily eat it raw. He slashed a broad palm frond from a tree then carefully cut slits in the leaves with his bayonet, to allow air flow as he smashed the bird.

At six feet from their quarry, both he and Branch froze, studying the creature as it tremored, seeking to lie lifelessly and blend in with the carrion of the slough. J.W. moved in from behind, Branch from the flank, cautiously avoiding the deadfall. J.W. slid the palm frond into the air then violently whacked down.

"Got it!" J.W. screamed and then lunged on top of the bird. The animal mustered no fight. "Baddest hunter in the swamp!" When Ranger Weathersby slid his hands under its belly, he detected a squishy feel, a coating of slime on the bird's underside. "Damn thing's so scared of me, it pooped itself."

As he grabbed the vibrating animal's neck, ready to twist round and round, the body turned belly up, and J.W. hurled it into the water. He gagged. Branch ran to survey their capture but choked at the animal's maggot-infested belly.

A mammoth, orange sun cracked over the flat horizon. The three drifted together for a map check, and J.W. asked. "Where you been all night?" Branch and Bearchild looked at each other and shrugged.

They had to listen for a while before hearing the platoon, and made their way back believing the quiet an artifact of sound absorption in the swamp, or perhaps, the men were sleeping, or best, wolfing down Cs. Approaching a clearing in the savannah, they stopped short as they spied soldiers dispersed, belly down, in a defensive perimeter, their M-14s poking out at the ready. A few steps closer, they realized not one of the sentinels was awake, and that there were empty C-ration cans and cardboard C-ration boxes sprinkled about the clearing. They ran unchallenged into the center of the tactical formation and reported in. "No one's anywhere near us. When do *we* eat?"

The cadre went to a lump in the savannah, lifted a poncho, and handed them each a box of Cs. He told them to go sit by a tree and take a load off; but before they got there, a squad of wild men vaulted out of the tall grass and ran through the clearing, whooping, cursing, and firing blanks at the gasping bodies. They disappeared as fast as they'd materialized, and that was how fast the platoon leader was relieved of duty.

Bearchild was tapped on the head with the femur of a creature that had only partially been subjected to the process of decomposition. He designed a forced march across a new plain of elephant grass, promising a reward at the end of the mission—a hot meal if they slogged the prairie in less than three hours.

J.W. poured a can of Meat Balls with Beans in Tomato Sauce down his throat as he listened to the marching orders. He turned to Branch and whispered, "No way, man. First of all, we ain't gonna get there in three hours. Next, they just gave us a meal. Not gonna get another in the same day. That's not the way life works, man."

Branch was too busy scarfing down his Turkey Loaf to answer.

The platoon moved out. Branch and Weathersby took the rear. After a few minutes, they sat against a tree, ate the dessert from the Cs, and napped.

J.W. was the first to look up. He listened for the platoon and was not disappointed. When Branch opened his eyes, J.W. laughed, "Hey, man. I feel great, how 'bout you?"

Branch nodded.

J.W. leaned back and bubbled, "I feel like I spent my whole life in the bush, a transient. Nothin' wrong with not knowing where you're gonna be spending the night. Nice and warm down here in Florida, but once you're asleep, it's all the same. They ain't gonna feed us for a while, so I'm not gettin' all that excited about sprinting through that grass shit. You?"

"Man," Branch nodded peacefully, "I'm just headin' off after a day at the office. I drop my pack here or there, ain't no never mind. I love my canteen cup, man. Watch this." He took the toilet paper packet from the Cs and used it as kindling to build a little fire. He poured water into his canteen cup and held it over burning scraps of dried grass until the water was lukewarm. Into the cup he sprinkled, a grain at a time, the packet of powdered coffee and sugar from the C-ration accessory pack. He stirred with his finger, licked it, and brought the cup to his lips. His eyes relaxed, and a gentle smile came to his face. "This thing ain't never let me down."

He patted his fatigue shirt pocket to make sure the last of the Winstons were still there. He grunted, "I'm a rich man, but I'll be richer with a belly full 'a Ham and Limas. Come on, move your ass. Let's don't screw Bearchild, man."

Minutes later, they saw the platoon break into a sprint, so they took off after them. Thirty seconds later, they came to the edge of the savannah. The barracks of Field Seven were a hundred meters to their front.

The noon meal in the mess hall was hot, spicy, and so salty, it tasted as if it had been pickled in brine for several weeks—even the milk. They were given time to chow down, and the troops looked at each other suspiciously until Sergeant Perez stood on a chair. Then they gobbled at the usual rate, but Perez simply urged, "Rangers, slow down. You aren't animals. Add as much salt as possible." When several of the troops looked at him inquisitively, he added without smiling, "*Muy caliente*, hot as hell out there. Rangers, this is the Jungle Phase!"

They were afforded twelve minutes to visit the post exchange for cigarettes and candy. As J.W. made his way toward the PX, Sergeant Perez stopped him and challenged, "Ranger, where are you going? Report to the camp commandant."

After the initial blast of adrenalin coursed through J.W.'s chest, he nodded to himself and airborne-shuffled toward the command shack, adding up the coming bonus points for having saved Fricker's life.

Headquarters Building was surrounded by a plot of white sand, a small area of which was enclosed by a heavy chain link fence. Free to roam the compound was a one-eyed, twelve-foot alligator. The reptile sat motionlessly, jaws spread wide, its chipped, ragged rows of green and yellow teeth a mountain range that dwarfed the Smokies.

J.W. muttered, "At least this CO has better dentition than Vock." A sign above the cage read:

BIG JOHN, THE LONGEST-LIVING RESIDENT OF FIELD SEVEN, TRAGICALLY LOST HIS EYE IN A FIGHT WITH CLASS 61-4. SEVERAL OF THOSE RANGERS ARE MISSING MORE THAN THEIR VISION. IT IS UNFORTUNATE THAT THEY WILL NEVER SIRE SONS TO BECOME AIRBORNE RANGERS.

And, in fact, an ugly scar covered the alligator's sunken left eye. Swarms of flying bugs entered and exited the cavern of his mouth in endless procession, a zoological Grand Central Station.

The commanding officer was waiting for J.W inside Headquarters Building. He had parked himself at a pint-sized, olive drab field desk. A lopsided, mildewed portrait of LBJ hung on the wall behind him. The stale air was stirred by a cock-eyed ceiling fan.

J.W. stood at attention. When the officer mumbled, "At ease, Ranger," J.W. looked down at the major's nametag. It read Schwartz-something, but it was so long he did not have time to finish reading it before the officer blurted, "Ranger, got a call from Captain Vock. Says you threatened him. That so?"

"No, sir."

"Why would he say that if it wasn't true? Are you calling him a liar?"

"No, sir. He hates me."

"That's correct."

"He calls ahead to every hole in the wall I get sent to. Honest to God. I don't know why, sir."

"I don't either. He's planning to let you go through the course to the end. Even the last problem. Then he's gonna stick it to you."

"Can he do that, sir?"

"He can try, Ranger. Better keep your nose clean. By the way, you know anything about that trouble out there last night? You were on point, yes?"

"No, sir, I mean yes, sir, I was on point, but I don't know anything about trouble. We just didn't believe it tactically correct to obey unknown voices in the swamp, sir. We've learned something in this school, sir."

He studied J.W. intently for a few seconds then muttered, "Dismissed."

As J.W. waited for his salute to be returned, the major added, "Sergeant Perez doesn't like to find Rangers with their pants down out on patrol. You're ridin' on the edge, Ranger. Outta here."

J.W. crept past the first sergeant's empty desk. His eye caught an abandoned P-38 can opener sitting on a pile of papers, and he slipped it into his palm. He stood on the porch to watch the cadres gather, preparing for formation. He cursed Vock for having kept him from replenishing his supply of candy at the PX. Morelotto, pockets stuffed, stopped by headquarters to humor himself by making noises at Big John. He tossed a couple of Mars Bars into the leviathan's mouth and laughed, "Stupid thing's got his bad eye turned toward me. Probably thinks it's hailing candy. They really are brainless fuckers, aren't they?"

The whistle blew, and as J.W. jogged toward formation, there was a terrible crash at the animal pen. Morelotto had tossed a rock into Big John's mouth then waited a few seconds to see what would happen—nothing, so he took a step closer. The slam of jaws sprayed a foul discharge into Morelotto's eyes and mouth. The young Ranger spit madly and blasted away from the compound at a sprint, begging Sergeant Perez, "Sir, is that shit from the crocodile's mouth dangerous?"

"I'm not a sir. Have you learned nothing? And yes, it is poison. Blind a man. Takes twenty-four hours. So you still a got a day to go. Now get your ass into formation."

Morelotto fell in but took one of his canteens, poured the contents into his eyes, and emptied the other into his mouth, gargling and spitting out the entire quart.

Following an inspection in which their just-purchased candy was confiscated, but cigarettes ignored, the company departed Field Seven and headed away from the swamp, soon entering a vast, dry patch of sand that stretched for miles. The salty earth supported only parched brown scrub, scrawny plants shielded with shiny, pointed thistles. J.W. consumed the last of his water at 1400 hours. He sweat that away by 1430.

Outbound a few hours on that mission, J.W. began combing old Ranger garbage tips for food. His forays into the world of the hunter-gatherer continued to be stained by failure after failure. He turned to Gillette moaning, "Man, you need a lotta hours to find something to eat. I'm wastin' too much time on cadres and tactical problems."

Gillette laughed, "To survive by foraging is a twenty-four-hour-a-day pursuit. There's an excellent treatise on the topic. It's a PhD thesis by one of my colleagues. His argument is that's why the original human hunter-gatherers didn't have time to develop skills like reading and writing, or even time to formulate new ideas like agriculture. Everybody was out all day scavenging. It was a Catch-22."

"Uh huh."

The less intrepid Rangers had given gave up searching, accepting they would suffer hunger and deprivation regardless of what they did to improve their lot in life. That left them a simple chore—put one foot in front of the other, and let the mind sleep. They were the large fraction of the class who believed the cadres' declaration that no Ranger had ever perished for lack of food.

J.W. studied his fellows. Most were sufficiently alive to have spent the energy to ease their load by emptying rucksacks of even a single pair of extra fatigues, a sleeping bag, or underwear. Canteens were kept only quarter full—in Florida, there was warmth and a plethora of water. The Jungle Phase was water.

When the need to replenish fluids came in normal times, a Ranger dipped his canteen into a pond and dropped in a tiny, off-white, water purification tablet. J.W. laughed that both the brown, wax-covered bottles and the pills themselves were indistinguishable from the nitroglycerin his Uncle Nat had used in fits of squeezing chest pain. Nat clutched his mini-bottle of heart-soothing medication in quivering palms all day, unwilling to be away from the life-giving medication, even in the men's room. As far as J.W. was concerned, his bottle of water-sanitizing tablets was as consequential as his uncle's heart-sparing pills. While the bottle added a few grams of extra weight, he appreciated the minuscule tablets which, when added to a canteen of river water, rendered a medicinal-tasting, but ostensibly microbe-free, brew. The beauty of sterilized river liquid was the added protein of whatever carrion had floated into the waterway. It was wet; it had calories—it did the job.

Their mission that afternoon was to attack aggressors in the dunes of Eglin, many miles north, though still within the bounds of the air base. J.W. perused the map for the next river or tributary, a spot to fill his canteens, but there was nothing except barren, dry plains for forty miles. By 1800 hours, his face was so dry and his lips so cracked, he dismantled his M-14 and scooped grease out of the mechanism to smear on his lips. Pangs of hunger slowly disappeared, replaced by an obsession to drink—anything.

J.W. planned to dip his canteens into whatever they came to first, as long as it wasn't a slough. There had to be a pond, one that had missed the mapmaker's attention. But the land only became drier with each sandy kilometer. As the company marched north, away from the Gulf of Mexico, Branch grumbled, "This is like the gotdamn Sahara, in the Gobi fuckin' Desert."

J.W. added through a parched throat, "Yeah, but there aren't any mountains in the Jungle Phase. This is what we dreamed about in the mountains. Remember? Flat and dry?"

"Shut the fuck up."

Worse than the dust of each step that blew into the air and settled on the thickened saliva coating his teeth was the plague of sandy grittiness that had infiltrated what was left of his clothing. He trudged along in silence, hour after hour, for moving his bleeding lips to complain or pray was laced with pain. Eventually, sand permeated J.W.'s boots, and each step seared the bottoms of his feet. An hour later, he limped so antalgically, his left knee ached as though that leg had shortened.

J.W. realized he had begun to hallucinate again, and he waited for the roaring voices. There may have been a few whispers, but he could not hear them through the pain that was building in his left heel. His foot burned as if stabbed by twenty tiny daggers. By late evening, the moonlit silhouettes wandering across the desert became walking trees. He tried to catch up with one to sit against it and put padding in his boot. When he couldn't overtake a single one, he let his shoulders droop, gathered what saliva he could, and declared in a gravelly whisper, "I quit this shit."

No truck or helicopter appeared to snatch him from the school, so he dropped to his knees to take the weight off his foot. "Screw 'em," he grumbled. "Not going any farther. Fuckers can come to me."

With the trees nearly out of sight, though, he sacrificed a drop of gun oil to coat his heel. As the boot came off, his hand brushed needles protruding through the inside sole. In the flicker of what was left of his flashlight batteries, he did an inspection. There was no heel. It had fallen off hours before, though he had no recollection of the event. The twenty tiny nails that had held it in place had pushed slowly into the boot, and over the past miles, into the flesh of his heel.

On one hand, J.W. was relieved to discover the source of his agony, and that he was not as demented as he had feared; though on the other, he was now faced with an impossible circumstance.

He limped, barefoot, across the wasteland in search of any human being who had the power to address his pain, even a lane grader or a cadre. Passing Branch, who was curled on the sand in a fetal ball, J.W. looked down and waited until the man acknowledged his presence.

J.W. asked pleadingly, "You got an extra boot?"

Branch stared up. "Man, what the hell you talkin' bout? Yeah, I'll give you a boot, for a quart of fuckin' ice water."

J.W. trod on toward the only animated shape on the dark horizon, a sergeant brewing coffee. J.W. advised the instructor of his footwear dilemma, and the man duly recorded the incident in his spiral notebook, mumbling that he would radio main camp and have J.W.'s other boots sent on the morning chopper.

"In the meantime, Sergeant, do you happen to have a pair of pliers so I can pull the nails out of my boot?"

The lane grader patted his pockets. "Sorry, fresh out."

"Well, then, could I get a little water to hold me over until we meet up with the mess crew?"

"Ranger, I'm not a deuce-and-a-half supply truck. You had a chance to fill your canteens just like I did. But you didn't, did you? Tried to put one over on us by travelin' light, huh?"

"Yeah, but Sergeant, you got four canteens."

"Yeah, Ranger, and you got four seconds to get your sorry ass back to your platoon."

J.W. retreated in a half-crawl past Branch, who looked up with only one eye. "You got pliers?" J.W. asked, but when Branch closed that eye, J.W. chose a patch of sand to sit and pull the nails out with his teeth. He chipped another incisor.

The platoon of Rangers did not reach the enemy until 0300 hours. They did not fire their weapons or even make pop-pop sounds, barely able to remain upright as they staggered in circles through the aggressor camp, so many zombies babbling about water. A cadre watched for a bit, shook his head, and ordered them to take seats in the sand.

A greying E-9, a command sergeant major, appeared out of the desert to take a position in front of the formation. He wore a Ranger Tab and jump wings upon jump wings above a Combat Infantryman's Badge embroidered with two stars over the musket— the third award of the decoration.

He began softly, "Rangers, you are tired, hot, cold, hungry, and thirstier than you have ever been in your lives, and probably more dehydrated now than you will ever be again. You don't want to fight. You want to drink; and if you don't get water, RIGHT NOW, you probably won't care if you live or die. Am I right?"

Those still conscious nodded, barely.

"Now, Rangers, I want you to do something. I want you to think about your family, your mother, your wife, maybe your grandpa who fought in World War One. Some of you have kids. But you're not sure life's worth living and fighting even for them, right? You don't have the strength to go on.

"But here's what I want you to do. I want you to think about what you would do if you got a call that some cretin broke into your house and tried to rape your wife. Huh?" He paused and stared at the men then put his hands behind his back and searched the sky. Eyes began to open. He drew a deep breath. "I know what you would do. You would tell, not ask, the cadres that you were outta here. You'd sprint back to Field Seven, grab your stuff, and run into Panama City without stopping to get a mouthful of water. You wouldn't even know you were dying of thirst, would you?

"Look, you are Rangers. That is why you were chosen for this school, and why you made it this far. You would get up and fight like a dog if someone tried to hurt somebody you love.

"Now, I will tell you about the Rangers at Pointe du Hoc. That's in France. Any of you ever heard about the place?" The regulars' hands shot up. "I was there—D-Day, Dog Company, 2$^{nd}$ Ranger Battalion. I was a young whippersnapper corporal. My sergeant, Joe Devoli, never forget him, when we first tried to land, Krauts hit our boat, and we capsized. Freezing—Atlantic Ocean. Sergeant held this wounded Ranger, just another private, held him out of the drink for hours, until a British rescue ship hooked us out of the water.

"Then they sent us back to the beach and up the cliffs—straight up. Jerries were standing on the precipice firing down at us, defenseless men on ropes. But Fritz had his job to do, and so did we.

"Most of the men to my left and right were hit and fell. But Colonel Rudder just screamed, 'Drive on,' and, Gentlemen, we did. We crested the cliffs, the few of us still alive. You think you've lost a lot of men? We were ninety out of two-hundred-thirty-three when it was all said and done.

"Most of the Jerries were gone, run away, when we came to the ridge, but we engaged the Nazis that had the balls to stay behind, and we drove the bastards back to Berlin. Someday, Colonel Rudder'll get his due. They'll name something after him. Damn well should. That's who you men need to be like.

"Now, I want you to get off your rear ends and form up into platoons and do the attack again. You don't have to pretend you're protecting your family, because Gentlemen, you are. You are the tip of the spear. This school ain't for the faint of heart.

"Oh, yeah, and you wanna know why Sergeant Devoli kept that private alive for so long, nearly drowned doing it? I asked him. Said, 'I'll tell you why, young troop; Guy owed me five bucks.'"

Morelotto stood. He looked around at his colleagues and walked to the sergeant. "I never knew that, Sergeant Major. Thank you, sir. I'm ready."

The rest of the company came to their feet. "Drive ons" brewed in the arid sands, and the attack was rejoined. When it was over, the men looked about for the senior non-com. He was gone.

The next mission took them even further from Field Seven and water. Though J.W.'s limp worsened, he wished the foot hurt more, to divert the misery from his throat. While Branch and Bearchild were only fifty meters away, J.W. felt extraordinarily alone, convinced no one could be suffering as desperately as he. At the next stop, J.W. left the platoon, setting out to find water. He hadn't traversed fifty meters before he came to a blubbery mass lying atop a sand pile, a lump that barely groaned as J.W. stumbled and landed on top of it.

"Morelotto, what the hell are you doing out here? Hey, wait a minute. What's that in your hand?"

"It's mine. Leave me alone."

"Bullshit. We didn't get Cs yet. Where'd you get that, man? Steal it from a cadre?"

"What if I did?"

"Nothing wrong with it. I mean it's the right thing for a Ranger to do. But just tell me what it is, man." J.W. tugged the can out of Morelotto's hand. "Fruit Cocktail? Hey, gimme a sip."

"Kiss my ass."

"I'll trade you my next B-3 unit for one sip. Come on, man."

Morelotto laughed arrogantly. J.W. fattened the offer. "Okay, main meal, your selection."

Morelotto laughed louder and licked the top of the can around the tiny hole he had punctured in the lid.

"Shit, three smokes, a quarter pack of granulated sugar, and my toilet paper."

Morelotto sneered, "What the hell am I gonna do with toilet paper?" His eyes became deeply thoughtful, and he nodded almost imperceptibly, guardedly allowing J.W. to wet his lips on the olive drab can. J.W. sucked a few drops of the syrup into his mouth, but the liquid was sopped up by grit on his teeth.

J.W. handed the can back and whispered through scorched lips, "Man, you got a lotta discipline, what with that tiny hole. I'm impressed. I knew you were a good man, a good Ranger. But why not make it just a little bigger? Look, I got a P-38 here. I'll donate it to the cause, man. What'a you say?"

J.W. stared at the can for a couple of seconds before his torso erupted, and his hand snatched the tin. He spun away from Morelotto and sucked as if it were his last hope for life. A few stringy drops trickled only as far as the back of his tongue and were absorbed in the layers of sloughing skin and sand. As he struggled to pull a few final drops into his mouth, Morelotto put his hand out for the can and glared. Weathersby handed it back and walked away, head sagging.

Morelotto put the can on the ground behind him, thought for a moment, and managed a few raspy words. "Hey, Weathersby, you okay, man? You're limping."

"I'm okay. One of the cadres stole the heel off my boot while I was sleeping."

Morelotto considered that for a second and asked hesitatingly, "Hey, man, were you telling the truth when you said you were proud of me? I mean, a decent Ranger?"

In the ten seconds it took Morelotto to speak, the molasses in J.W.'s mouth was gone, evaporated, the throbbing now critical. But the old sergeant's image came to him, and he composed himself. "Man, I don't know. Most of the time you seem like you don't care. Then there's that flash of light. And here you are, one of the last standing. Kinda funny, don't you think? You and me still here. Must mean something, huh?"

Morelotto shook his head. "The only reason they didn't boot my ass outta here is on account of my father. Pure and simple."

"That's bullshit, man. I watched you. Most of the time you're right up there with the crowd. You done better than me a lot."

"Man, you think so?"

J.W. looked around and whispered, "Look, I been on the verge of quitting and getting my butt sent to the First Infantry Division more than anyone here. I was the wild card, and they hate me. I begged like a baby not to come. I was sent here for punishment. Everybody knows."

"Really? I didn't know. I thought you were one of the guys. Pals with Branch and Bearchild. They hate me."

"Well, look, you said some shit to them. You can't do that, man. This ain't the Brown Boot Army, General Patton smackin' people around, no more. If a man works hard, and he's smart, you gotta respect him. Am I right?"

Morelotto brightened for a moment. "Yeah, you are. You think it's too late?"

"Too late for what?"

"Walk outta here and shake hands with your buddies. Maybe they'll give me a hand when the shit hits the fan?"

"Fuckin' A. They're good guys. Smart as hell. Scared just like you and me."

"They sure don't seem scared."

"Where they grew up, they can't show that shit. You think you're the only one here with problems? Turns out not nobody's on top of this lunacy—not one."

"Yeah, well, they all seem to be doing better than me. And I don't have a choice. The rest of you do. If I fuck this one up, too, it'll be the last straw with my old man."

J.W. nodded. "You'll make it. I know you will. Just drive on, man."

He handed Morelotto a smoke, and the two puffed away, praying for a homeopathic cure, using the acrid smoke as a drying agent to elicit moisture. The experiment failed, so Morelotto opened the lid on the Fruit Cocktail with J.W.'s P-38, and the two wolfed down the soggy green grapes and pear cubes. It wasn't worth the caloric expenditure to have poked the can thirty-eight times to pry off the lid. J.W. pondered his life once again, embarrassed at his weeping in the previous phases over trivial matters like sleep deprivation and starvation.

When the sun rose, J.W. walked with Branch and Bearchild. He stopped and blurted, "You know, Morelotto's not such a bad guy."

Branch looked at him askance and harrumphed, "My ass. You got too much sun."

"No, man, why don't you give him a chance? He's as bad off as the rest of us. You can never tell how a man's gonna handle this pressure shit."

"Like I said, my ass."

Three minutes later, Morelotto was touched with a bouquet of brambles. The lane grader harped, "Last chance, Ranger. Fail three patrols and you're done. Got it?" The man lifted himself to sit on a fifty-five-gallon oil drum poking out of the sand. He watched for a bit as Morelotto stared, frozen, at the manila envelope. "What's wrong, Ranger? There's an objective, a stronghold in the dunes that's a waitin' for you to capture it."

"Gimme a minute to study the order."

"This ain't college. No studying allowed. You need to get movin', pronto like. This is for real. The enemy has invaded and taken

Miami. Their rear units have advanced to a point twenty-three kilometers east of here. Your orders are to attack them and at least save the panhandle of Florida. Do you think you can do that?" Morelotto stared at the papers. "You listening to me?"

Morelotto jumped up and pointed at the man. "Hey, you listen to *me*! I'll plan the damn attack, but I get the same amount of time as everybody else. I'm no different than the rest of the Rangers. I'm not gonna go runnin' off before I know what's goin' on. So cool your jets, *Sergeant*."

The man popped off the oil barrel like a cork out of hot champagne. He stiffened in front of Morelotto, fists clenched, but J.W. placed himself between the lane grader and his new pal. Then the other Rangers, half of whom had been dozing, jumped up and surrounded Morelotto, facing out, musk oxen protecting a calf. The cadre backed off, retook his place on the barrel, then keyed the mike on his radio.

It was finally over for Morelotto, and J.W. wondered who his father could be, and how bad it was going to get. It would not take long for a chopper to fly in to haul the latest casualty back to Field Seven. He was ashamed that he felt a tickle of excitement inside with the thought that the helicopter might bring water and a new boot.

Nothing, though, fell from the sky except the unabated rays of the tropical sun. Morelotto moved the platoon further into the dunes which became higher, the sands looser, and the brush drier, their prickers now honed razor-sharp. They marched directly away from main camp, not toward it, and as the day grew hotter, J.W. fretted that if a freight-laden helicopter did try to make an approach, it would be unable to land in the boiling, thin air.

Time slowed, each step lasting minutes. J.W. tried to pool spit in his mouth and save it for one good swallow, but it was too thick and granular to scrape off his teeth. Then he attempted the Boy Scout maneuver of sucking pebbles, but the tiny rocks added additional grit, no matter how hard J.W. buffed them against his fatigue pants. Assuming a larger pebble would be easier to hold while polishing with bleeding, infected fingers, J.W. burnished a chunk the size of a grape and placed it carefully in his mouth. A few drops of viscous saliva were expressed, and he nodded to himself at how proficient he had become at survival.

He was content until he tripped, fell, and swallowed the rock. For a moment, his stomach smiled at the offering, then cramped violently until he vomited. Another tooth chipped as the rock flew out.

He reminded himself what Miss Kreutzer, his music teacher in seventh grade, told the class. "Young ladies and young gentlemen, you must never forget that there is a silver lining in *every* cloud. It will serve you well in life."

In this storm, he assumed it was the thirst, and how it kept his mind off his missing boot heel. Eventually, however, the spasms in his calf hobbled him, and he pulled out of the column to hide behind a grove of yucca, cutting one of the broad leaves into a cookie for his boot. As he hacked at the frond, he noticed a thick, green liquor oozing from the cut surfaces. He smiled and called out, "Glory halleluiah! That's it! The silver lining!" The others turned toward the bush, but seeing nothing, accepted the proclamation as just one of their own, private voices.

J.W. licked a few drops of the astringent fluid, the bitterness provoking his already delicate stomach, propelling up acidy mucous laced with particles of dirt. He smiled, whispering under his breath that, on balance, the vomitus tasted better than the yucca sap.

At noon, the platoon reached the jump-off point for its attack on the enemy troops intent on capturing all of Florida. The brush had become heavier, and J.W. crouched behind a shrub to avail himself of the toilet facilities, wondering the whole time what could possibly be left in his body. He squatted near the earth and promptly fell asleep, waking minutes later to the vision of a cluster of black-bereted aggressors retreating into a wooded area a hundred meters off. J.W. watched with singular interest as the soldiers dropped their packs into a hole, which they camouflaged carefully with brush. Biding his time, biting his lip, fidgeting, he hid until the enemy dispersed, then low-crawled across the barrens to the cache.

Not wasting the time to strip brush from the hidden reserve, J.W. shot his arm inside, prospecting with his hand until he came to a cool, metal cylinder. A flash in his addled consciousness told him God wanted him to have that treasure, a chilly beer, and as J.W. pulled it from the crater, his hand closed on a button protruding from one end of the tube. It hissed like a green snake. "A can of Cool Whip!" he shouted. But the scent that followed the hiss was tart, and though distantly familiar, J.W. feared it might not be something edible. He was correct. As the can was drawn from the pit, the blasting sun reflected on a can of Right Guard. Accompanying a string of profane epithets, he pitched it over his shoulder then lunged his arm back into the hole to renew the mad groping.

J.W. remained oblivious to the world outside the pit until a not-unfamiliar sense of impending doom blanketed him. He shook it off at first, a mental reflex, he was sure, for having taken so much time to complete his mission. The sensation, nonetheless, grew deeper, and he took a slow, deep breath before turning his head. Spit-shined boots reflected a gaunt, sick face in the mirror-like toe.

Before J.W. looked up, he chuckled nervously, "We gotta stop meetin' like this."

"Ranger," the figure sermonized like a Southern Baptist minister desperate to salvage his congregation, "The good Lord put me on this Earth to watch over you and make you do right. I feel like a failure in the eyes of God, Ranger, and that ain't good."

J.W. babbled, "First Sergeant Cowsen, you see First Sergeant, I saw one of them green snakes shoot across the dunes into this hole. I was trying to catch it and return what a fellow Ranger had lost. I don't know if you heard about the incident, but it's really true."

"Ranger, how do you figure a snake came up with a can of Right Guard, carried it to its den, and covered it with brush? They don't have armpits to use it on, 'cause'n they don't got arms, and neither will you if you don't get off the ground and come to a position of attention. Anyway, the green snakes live in the swamp, and there ain't no water within twenty miles of where you're layin'.

"This may be the big one, Ranger. Bein' as you're AWOL, I am placing you under house arrest, and this is your house. You will stand guard over this hole, at attention, for the rest of the afternoon while your platoon lounges and drinks water, as much as they can put down."

"Water? First Sergeant!" J.W. howled plaintively, but Cowsen marched off through the brush, soon cresting the horizon and turning to point an angry finger at J.W. before disappearing beyond a dune. J.W. had his orders, and in lieu of fluids, he placed his Ranger hat on the highest branch of a thorn shrub and dropped for a nap. Though as hot as he had ever been, curiously, he did not sweat.

He dreamed again of Viet Nam and plodding through a steaming jungle, searching for an enemy that was out of ammo and throwing rocks instead of firing lead. In the nightmare, one of the stones hit his head, and J.W. awoke with a start. He jumped up, bracing at attention until he realized he was alone, the victim of another painful fantasy. He lay back down, but before he could fall asleep, he thought he felt another missile strike his head. He didn't bother to look up and was alarmed that his hallucinations were now becoming painful. At least the voices were silent. A moment later, they returned.

"On your feet. Sleeping on guard duty is punishable by death in the Uniform Code of Military Justice, Ranger. Prepare to meet your maker."

J.W. was on his feet, blinded by the sun, but slowly able to make out the form of a man in an intact pair of fatigues. Then Weathersby focused on a set of sergeant's stripes. J.W. had no trouble seeing the man remove a .45 from its holster and drag back the slide. Arm extended, he pointed the weapon into the cloudless sky, slowly bringing it down into firing position. J.W. looked up at the barrel, though his eyes fixed beyond on the weapon on the perfect, cloudless sky. His was surprised a man could die in such peace and quiet.

"Good. I don't give a shit, Sergeant. Send me back for court martial. Kill me, I don't care. I qui…"

As J.W. formed the last word, he was distracted by the whap of a Huey main rotor roaring in on a breakneck approach. J.W. could make out the pilots grinning fiendishly as they skimmed so low and fast, the sergeant threw himself to the ground. The ship skidded to a sand-enveloped landing fifty feet away, and J.W. abandoned his own execution to sprint toward it, relinquishing the last of his energy on what he believed to be his final run in Ranger School.

Fresh cadres spewed from the belly of the ship, and J.W. looked past them for jerry cans of water. The first officer off, however, chewed J.W.'s ass and pointed him back to the platoon.

J.W. stood fast. "Sir, where are my boots?"

"They're on your feet, Ranger. What is your problem?"

J.W. looked down. Indeed, his boots were where they had been for the past months. J.W. nodded. "But you see, sir…"

Then J.W. caught sight of the sergeant who had been readying to consummate J.W.'s punishment. He was running toward the aircraft, fists tight, one still gripping his .45. J.W. slunk to the other side of the helicopter. The sergeant jumped aboard the aircraft and made for the pilots. J.W. could see heads bobbing about the cockpit, locking their stares on him.

A newly arrived first lieutenant hopped off and raised a thumb. The ship crept forward on its skids for takeoff.

With the helicopter gone, the fresh lane grader turned back to J.W. and lashed heatedly, "Get back to your unit, Ranger, before I tell 'em you're AWOL and have you executed."

As J.W. hobbled to the main body, Morelotto ran to him. "Did you get any water, man? Could you share just a little? I gave you some of the Peaches in Syrup. You owe me."

J.W. snarled. "What in the hell water are you talkin' 'bout? You got water, I didn't, fool. And it was Fruit Cocktail."

"Cowsen said you stole water and were executed for it," Morelotto spat back.

"No, man, he told me you guys were drinking all you want."

Morelotto started to grab at J.W.'s canteens, but the lane grader ran over and ordered the two men to move apart fifty feet, then sent J.W. to the rear for security detail with Branch and Bearchild. "I hear you're the only ones who are reliable."

As the platoon departed, J.W. laughed to Branch, "Pretty damn sad that they want to execute me, twice in ten minutes, and I'm the one they trust."

The platoon stopped at sunset. A supply helicopter rumbled in, fostering in J.W. and his mates smiles and, "Thank Gods."

There in front of them was the water and food for which they had waited patiently, and despite the dust storm generated by the main rotor, the entire platoon drifted toward the bird, foot over unsteady foot, columns of the trotting dead. But only a couple of fresh cadres belched from the belly of the dust-engulfed helicopter, their canvas canteen holders dripping wet to allow evaporation and cooling of the water.

On the other hand, J.W. took the godsend of new lane graders, and the twenty minutes it would take them to figure out who they were assigned to torment, to creep away in search of a puddle. He shuffled dolefully through the infinite sand, cresting a particularly high dune to peer over its summit into an abandoned garbage tip, rusted C-ration cans and rotted cardboard strewn over half-an-acre. The very sight of food containers sent J.W.'s salivary glands into painful contraction, and he prepared to wet his lips with the coming saliva, though none came.

"No problem," he mumbled, "I'll get somethin' from this trash to trade a cadre for water." He inspected each discarded can carefully, finding them all bone-dry, save for the predictable crust in the Ham and Limas. He piled those tins into his pockets to entice Branch into a future negotiation for water. Some of the Cs had been opened on the bottom with small slits, the food sucked out by the enemy, packed with sand, and then placed right side up to look and feel as if they were full.

Though it was a fruitless pursuit, the speech area of his brain chattered on, while another section of white matter tallied the calories already squandered seeking food. No matter, his heart would not let him stop, and he dutifully booted each can, including a squat B-3 Unit which had heft but was surely a ruse. Something, however, about the way that tin arced piqued his interest. Its trajectory was distinct, a trace higher than the parabola of the sand-filled species, and much lower than the path of an empty. When it landed, it was an inch or two short of the spot his subconscious had come to expect. He cursed himself yet again for having slept through physics classes. He made the resolution to walk up to it and kick it again. When no sand sprinkled from either end, he took the next step in the food-search algorithm—he bent over, investing sufficient energy to lift the possible treasure out of the dirt.

The label read Fruitcake. J.W.'s heart skipped a beat. Slowly, tediously, deliberately, cautiously, he checked the top—intact, as presumed. He turned the can upside-down—the bottom was pristine. He examined the metal for pinholes in the side, for perhaps the cadres or aggressors had injected poison. Negative.

His heart pounded in wild expectation. Was it really happening? Was he about to score? J.W. calmed himself and took a seat amid the garbage to savor the moment of good fortune. Slowly, with reverence, he placed his hand in his pocket for the P-38 can opener he had purloined from the First Sergeant's desk. He imagined bringing the tip of the opener to the edge of the can, then investing the energy to make those 38 revolutions, but the pockets still attached to his fatigues were empty. He walked in circles, frantic, searching every inch of the garbage pit before he recalled trading it for a few drops of Morelotto's filched Fruit Cocktail.

He went for his bayonet. "Fuck, I'll stab the thing open." As he lifted his M-14 over his head as he had been tutored on Day One, a visual hallucination stopped him in mid-thrust. This time it was an image of his Ranger brothers, and then the voice of the old sergeant. He kicked jauntily back to the platoon, grinning for the first time in weeks, singing reverently, "Seek and thee shall find, brother. And the Lord has found this poor sheep."

Branch looked up and shook his head. "What the hell you mumbling 'bout?" He turned to the others, "Looks like this one's done for."

"No, no, my brother, it's not the end. It's the beginning. I've found grace and mercy, baby." He pulled the cans of dried Ham and Lima crust from his fatigue pockets and shook them in Branch's face. "For you, I bring heavenly gifts."

Branch's eyes reddened, but his body was devoid of the fluid it would have taken to produce tears.

J.W. wiggled the can of Fruitcake in several of the men's faces. Though their pupils narrowed to pinpoints, not a single hand shot forward. Even the liberal arts majors calculated the wasted caloric expenditure of reaching for a phantasm.

J.W. held the can toward the heavens and boogied. As he hopped about the dunes, Morelotto pulled out the P-38. He crawled forward and held it forth for J.W., a disciple at his lord's feet.

"Now, gentlemen," J.W. went on, "please watch as God's chosen person samples the fruit of the loom."

"A slice?" Morelotto pleaded. "I shared my Fruit Cocktail with you this morning."

Branch snapped into sitting position. "Fruit Cocktail? Where the hell did you two chumps get Fruit Cocktail?" He came to his feet and took an angry step toward Morelotto, who puffed his chest, the muscles of his face tightening.

Branch reached out to grab Morelotto by the lapels of his fatigue jacket, but Bearchild rose. "Both of you, that's enough. Everybody gets a small chunk." He snatched the can out of J.W.'s hand and opened it but then handed it back to J.W. "Slice it up. Twenty-three shares. All the same."

"That was my plan the whole time."

"Uh huh."

As Second Platoon sat sucking on dry morsels of sugared fruit, expecting saliva to dissolve the bone-dry manna, Branch jabbed his slice forward touching Gillette's. "Compliments to the chef," Branch toasted as he nibbled with the greatest of moderation on J.W.'s tissue-thin offering.

Gillette wondered if there was a retail outlet for the product, proposing, "I'm going to purchase cases of this item when I get home. Send it as Christmas gifts to my dearest friends. This is a real treat. Bravo, Weathersby."

It was not long, however, before the Rangers of Second Platoon were making faces, scraping their withered lips and tongues against gritty teeth.

"Man, that shit was dry," Morelotto bitched. "Now, I'm even more thirsty, but screw it. Thanks Weathersby. You're a good Ranger."

Several of the men called out, "Yeah, man, thanks." Morelotto looked over at J.W. and beamed.

J.W. gravitated toward the rear, where Branch and Bearchild had taken up residence. Though Bearchild rarely smoked, he took a cigarette, holding it awkwardly, allowing the smoke to filter through the scrub pine into the mauve sunset. He sat cross-legged, as if in prayer, the smoke and the sky creating an ethereal forum.

J.W. was first to speak. "I figured it out. I'm going to be a doctor someday. A doctor doesn't starve. I never heard of a starving doctor. He can trade a shot of penicillin for a chicken. I'll never starve again. That, I guarantee you."

Branch laughed weakly, "Man, you got three more years on active duty, a year or two in Viet Nam flyin' round in a HUEY death bucket. Pilots is a dyin' as fast as they sends 'em there. If'n you live, it's another year or two of premed when you get out, then four years in medical school and three more in residency. And then you can start a practice. That's, let's see, that's eleven years to go before you start makin' a living. That's crazy. You'll be near death by that age, if you manage to live that long."

"Thanks for the support, man. Don't you worry. I'll put the years of work in," J.W. said, setting his jaw. "I got no problem with time. But I promise you, by thirty-five, I'm going to be in charge. No one's ever going to do this to me again. When I came here, I didn't know what in the hell I was going to do with my life. I do now. Thank you, cadres."

"Doctoring's easy living. That what you figure?" Bearchild asked sarcastically, "Deciding who lives and who dies?"

Bearchild's face hardened, and he turned away. Though he was as quiet as usual, there was a sullenness neither Branch nor J.W. had previously witnessed. They dared not challenge him.

Turning to J.W., he conceded, "You're okay, man. Just promise you won't turn away the poor."

He became quiet. Branch and J.W. assumed Bearchild had finished one of his most protracted discourses in Ranger School, so they leaned back to sleep, but Bearchild monotoned, "When my brother was little, he got sick as hell. None of the doctors in town would see him. They told my mother to take him back to the res, to the DIA clinic."

"What's DIA?"

"Department of Indian Affairs. That's where we had to go. It was closed for the Fourth of July. So she drove back to town to try again. Then the car broke down in the heat. I was holding him in my arms. He was hotter than the tar on the road. When he stopped moving, I thought he was dead. So did my mother. So she started to cry, and she said it was a curse to be born Indian." Bearchild stopped.

J.W. lit another smoke. "What happened?"

"Steven War Bonnet? He had meningitis. He lives in a home on the res now. He just moans and cries, twenty-four-hours-a-day. He has seizures every few hours. Mostly he lies on his back, screaming and whining, whining and screaming."

"Does your mother take care of him?"

"Nah, she died a few months after he got sick. My father cared for Steven for a while. I guess it was better Steven was taken away. My father blamed himself. Then I was the only thing he had left. Embarrassed the hell outta me when he started going around town tellin' anybody who'd listen that his son was gonna be the first Rose Bud Sioux in the world to graduate from Harvard and then go to Congress and change the rules."

Branch nodded. "And so you graduated from Harvard. Congress next?"

"Who knows. Gotta live through the Nam first."

⟞╀ ╀⟝

The most recent platoon leader, Gillette, who had transformed himself into a great Vietnamese general at Harmony Church, now considered himself an Erwin Rommel, a latter Twentieth Century Desert Fox. He laid plans for Second Platoon to attack the next objective by trooping straight across the sand and fooling the aggressors. In his briefing, he crowed that the enemy would assume the Rangers would take the easiest course, the one in the river bed the map proclaimed was dry during the winter months.

Morelotto interrupted. "Hey, man, what plans have you made to resupply? Particularly water. You got a responsibility to care for your men, ya know."

Gillette ignored him, consumed with defending the impossibly complicated assault he was explaining to the cadre. Hoping to get some sleep, the cadre urged Gillette to shorten the raid. During the squabble, J.W. took a moment to peruse his own map, searching for a tiny creek. He detected a small finger of slough obtruding into the sand, not half-a-mile off the course chosen by the Desert Fox. J.W. suggested they divert to fill canteens, but Gillette groused, "That's poison. And anyway, we're not wussies, we're Rangers."

Branch, Bearchild, and J.W., sent to man positions on rear patrol, waited until the unit departed then sprinted for the water. As the sun dwindled, it became an effort to see obstacles in the flat light. Nonetheless, they pushed on, driven by a worsening thirst that caused them to collapse every hundred yards or so and lie prostrate on the dunes moaning. On one of the more theatric disintegrations, J.W. tumbled into a bog, coming to rest with slime just millimeters from his face. It was all he could do not to suck in quarts of the rancid fluid that was so close to his burned lips, but the stench provoked within him an instinct not to consume obvious poison. Instead, he filled his canteens, sieving through his fingers sugar-cube-sized remnants of animals that had migrated to the slough to die. When he was done, Branch shined a light into the swamp, and J.W. could see vestiges of carcasses, one of a rotting porcupine, washing back and forth. The carrion swayed gently in the ripples generated as he crawled out of the stinking liquid.

J.W. could not find his water purification tablets. He cursed that an aggressor had stolen them while he slept. Branch at first refused, but then, reminded of the crusts of Ham and Limas, and the sliver of fruitcake, relinquished two pills. J.W. dropped one pill into each canteen, shook the stew, and put the containers back on his web belt to wait the requisite twenty minutes for the medication to work. With the sound of sloshing coming from his canteens, though, he controlled himself for forty seconds before ripping them out of their canvas holders to chug-a-lug both.

The water, lumpish as curdled sour milk, manifested its toxicity in seconds, forcing J.W. to vomit. "Shit," he cried, "that shit's worse comin' back up than goin' down."

The three rejoined the platoon, J.W. periodically pausing to gag gummy vomitus and particles of animal. Each time he hocked the waste, a squirt of the familiar green liquor burped into his mouth from his innards, continuing the never-ending, vicious cycle that had begun years, he believed, but only forty-some days before.

At midnight, the platoon traversed a narrow, freshwater stream. The lane grader and several cadres chased behind, yelling into the ears of every Ranger, "You are not to drink this water. It has been poisoned by the enemy. You will die within minutes if it touches your lips, Rangers."

In the dark, however, it was easy to hear the plopping of Ranger boots, then mad splashing as if the opening day at a new swimming pool, and finally the gurgle of water bubbling into canteens.

By dawn, some claimed there had been an attack, and that they had beaten back the enemy, but J.W. recalled nothing of the night except the tempering of the dreadful taste in his mouth. As the aggressors had withdrawn into the jungle, the Rangers' next mission would be to ferret out the survivors.

The entire company gathered deep in the swamp to finish off the invaders. Intelligence revealed the enemy had fled in rubber rafts, and the jungle was too thick for Rangers to traverse on foot.

A cadre briefed them, "Rangers, you are about to experience the essence, the jewel, of jungle school, an amphibious assault. The exercise will begin with a short walk south along Turtle Creek to its confluence with Live Oak Creek, near the town of Mary Esther. Gentlemen, the embarkation point is on the Gulf of Mexico. Tactical orders direct you to man rubber rafts hidden by partisans. You will navigate inland rivers to find, fix, and destroy the enemy in their stronghold.

"Secret documents held by the enemy on that island are vital to the security and future of the United States. You may have seen the cherished portrait of our Commander in Chief, President Lyndon Baines Johnson. He, himself, has commissioned this company to save the nation."

J.W. raised his hand at the end of the briefing and asked, "We gonna be awarded The Presidential Unit Citation if we win?"

"Yeah, that's right, Ranger. And I'll be there to pin it on your ass."

When J.W. spied the first partisan, he flinched and slunk to the end of the line, but the man invited that very fire team into his rubber boat, smiling broadly as he handed out paddles. He took one for himself from a different pile. It had a blade nearly twice the size of the others. He made paddling motions with his arms as if loosening up, but as they shoved off, he dropped the paddle on the rubber deck under his feet. His efforts were limited to prodding his galley slaves to row more vigorously against the stiff current of the Yellow River. He stared with polychromatic eyes at J.W. and directed most of his antagonism toward Weathersby.

For the first minutes, they enjoyed being off their feet, but arms soon tired, then burned. The partisan at the helm relented, offering, "Go ahead, take a break, Rangers, but wait until none of the other cadres are looking."

So Second Fire Team of Second Squad of Second Platoon paddled slowly until the last of their mates' rafts sailed beyond a thickly vegetated bend in the Yellow. Within a millisecond, J.W.'s crew ceased digging at the water. They swore to the partisan, "It'll only be for a minute, Sergeant."

They drifted gently, and J.W. fell into a heavenly sleep, awakening as the boat orbited in a serene whirlpool. He was quietly at peace with the world until the sergeant pointed at J.W. and ordered, "Do a map check, boy."

"Ah, shit," J.W. cursed, "the current's pushed us downstream. We've floated past where we began at Mary Esther."

"Rangers," the instructor warned, "this will reinforce another Ranger commandment: circumvention, like crime, never pays. Now, get paddlin' to make up the time you lost."

As they thrashed upstream, J.W.'s efforts fell short, according to the partisan, who ordered J.W. to commence twenty-five-and-one on the rubber gunnel. The sergeant rasped, "You," he pointed at Morelotto, "put that Ranger's paddle on his back so we can be sure he's doing perfectly balanced exercises and doesn't tip us over."

It took only three or four seconds before the sergeant was rocking back and forth as he sang an idiotic tune. Five seconds later, J.W. lost his balance and fell into the river. The paddle was carried away with the current. "Now, you've done it, Ranger. You're gonna pay for the missing equipment, *and* you're doing the paperwork."

The worst that could happen, though, was that he'd have to share a paddle, and all his mates would get periodic breaks. That made J.W. smile. The sergeant smiled as well as he thrust the paddle with the blade as wide as an airplane wing into J.W.'s hands. Every time J.W. dipped his oar, the boat yawed to the other side.

Back en route, they passed a string of rubber rafts lazily turning circles, snoozing crews overseen by lounging cadres who puffed blithely on cigarettes. J.W. called out to wake them, but that good deed begot fifty-and-one with hands and feet back on the rubber gunnel. This time, the sergeant gave Morelotto twenty-five-and-one when he tried to put the colossal paddle on J.W.'s back.

The assault came to pass, as did all things, or so said the Ranger School devout. There were no stray Cs this time, the aggressors having hoisted their comestible-loaded ponchos into trees. The cadres warned, "Rangers, those sacks contain deadly serpents, like the green one you idiots lost. Anybody caught messing with that stuff will be considered attempting suicide and will be executed. In the Uniform Code of Military Justice, it is a capital crime to commit suicide. You cannot graduate from Ranger School if you are dead, but the UCMJ does not have a clause that says you can't repeat the jungle phase if you are not alive."

Gillette queried, "Sergeant, excuse me, but that's like a triple negative. Could you please rephrase that threat?"

The platoon was sent to the left, onto a patch of broiling sand, to work off their sentence—Gillette was ordered right, to do them under a tree. The instructor laughed, "Sun musta got to that one. He needs to be in the shade."

The enemy successfully joined, Second Platoon reboarded their boats, hungrily anticipating the reward, a float with the current downstream. The new orders mandated they accomplish the next mission and be back in the Gulf of Mexico by 0900 hours. That translated to paddling harder and faster downstream than they had up. There was a rider attached to the orders—a promise that they would be transported via motorized launch, and then truck, back to Field Seven for breakfast, but only if they destroyed the adversary on schedule.

There were cheers, but Gillette interjected in a whisper, "I think my hearing's returning. I think he said an 'on-time attack.' Gentlemen, that is an oxymoron."

The trip was uplifting, cadence sung for the next hour-and-a-half by the seven dozen remaining Rangers. Morelotto howled the lead:

> I want to live the life of danger,
> I want to be an Airborne Ranger.
> Someday my son will be like me,
> He'll run all day,
> He'll jump for his pay,
> Airborne,
> Ranger,
> Airborne,
> Ranger!

No one noticed when 0900 came and went, a milestone that passed with the troops still pulling at the river, singing happily. At the mouth of the Gulf, the company boarded motorized launches, as promised, and the next few moments were amongst the most peaceful of J.W.'s life. He reclined, taking in the early morning, tropical sun on the calm ocean, translating across the face of the Earth by dint of labor other than his own. The partisan at the helm quietly smoked and steered the Mercury outboard, a single bead of sweat hanging from the tip of his nose. J.W. felt pity for the man forced to keep his eyes open while the Rangers slept.

The silence of a failed engine, though, startled them awake. "Rangers, disembark."

Morelotto jumped up in disbelief. "No breakfast?" He screeched at the helmsman. "I demand to know why. You people are fuckin' liars."

The sergeant glanced conspicuously at his watch. "Gentlemen, you did not fulfill your part of the bargain. Leave your packs, web belts, and boots aboard this watercraft and swim the short mile to the beach. Your gear will be delivered to the few of you who make it, those not consumed by gators, puff adders, or jelly fish."

J.W. stripped off his fatigues, rolled them into a ball, and stashed them in his crotch, converting himself into a streamlined torpedo. His legs porpoised for fifteen minutes. When he arrived on the white sands of the Gulf of Mexico, he laughed at the splashing Rangers still a half-mile out. As he searched for a shady spot to drop and sleep, he stepped on a sideways-scampering crab, which expired in a jumble of guts and seaweed. J.W. took the remains into the bushes and ate them.

Beyond the beach was a thicket of yucca. J.W. crawled into it for an after dinner siesta. He spread his wet fatigues over the top of the spiked leaves and closed his eyes, lying in the nude and savoring the pounding surf. The last thing he remembered was babbling, "It's like a tropical lullaby."

When he woke in the shade of his freshly-laundered uniform, he remarked to himself just how fast things dried in the tropics. As he dressed, he caught sight of Branch and Bearchild, sitting cross-legged on the beach, flanking his gear, staring out at the water. As J.W. approached, they looked up from their vigil, eyes widening.

"Man," Branch declared with disbelief, "Where the hell you been? We thought you drowned at sea. They said you were the only Ranger ever to die on the long swim. Been plenty died at Boiling, Creek but not in the Gulf. Even Cowsen had a few words."

"He's here? What'd he say?"

"He said, 'And of all Rangers, the one who could swim forever.'"

"No shit! Did he seem upset?"

"Man, I don't know. He didn't cry, if that's what you mean. But Bearchild, my man here, he bawled like a little baby."

That brought a bare grin to Bearchild's face, and Branch lifted a Chesterfield toward J.W., then a light; Bearchild took one as well, and the three sat quietly for a few minutes, watching several small launches circle in a searching pattern a mile out to sea.

When the platoon was called into formation, Sergeant Perez did not smile. "Shit, you're alive? Ranger, you will swim out to the search boats and tell them your body has been recovered. This time you will wear your uniform, and I'm going to tie your boots together. Get goin'! And the rest of you mourners, no breakfast until he gets back."

At the launch, J.W. sputtered, "Ranger Weathersby reporting."

The pilot, without looking down into the water, ordered, "Into the boat." The cadre was silent until they reached shore, but as J.W. disembarked, the sergeant commanded, "Ranger, you had no right to accept a ride back in this boat. Just for that, you're gonna dry shave, and your buddies are waitin' here until you do."

J.W. was frightened of the sores that would reopen on his face, but that morning, his beard was so soft from the hours in the water, he relaxed and drew the rusted razor effortlessly over the silky stubble. "Ranger, you look like you are pleased with your penalty. You are hereby awarded fifty-and-one for enjoying yourself."

Then there were an additional twenty-six allotted for having disappeared during the swim, and twenty-six more when J.W. protested, "Sergeant Perez, you can't punish a man twice for the same crime. It's in the Constitution. Ask Gillette."

Gillette blanched at the mention of his name, but his color returned during the push-ups he was dealt for quoting the Federalist Papers.

At Field Seven that night, J.W. complained of feeling fat and soft as he lounged in the mess hall after the meal. Morelotto used the hiatus to visit his friend, Big John, whose jaws were in their customary position, spread wide, motionless, bugs flitting past ruby lips, in and out of his drooling mouth. Morelotto tossed the requisite candy bars, knew that he had four or five seconds before the crash, so waited three and laughed that he still had a second or two to turn away. The flood of mucous came a tenth-of-a-second later, coating the side of Morelotto's face. He poured just a few ounces of water from one canteen but rubbed the skin red with his fatigue jacket.

With four and a wake up to go, the Ranger remnants were afforded a thirty-minute respite at the barracks to prepare for the final tactical mission. J.W. looked over the survivors, the gaunt, tanned, and numb readying themselves for the last maneuver, an easy one, it had been rumored. The closing days were a walk in the park, for the cadres had extracted their pound of flesh, and now all the soldiers were back on the same team, rebuilding physically and mentally so they might look and act like soldiers when they arrived in Viet Nam, a month hence.

Nonetheless, there sat an uncomfortable nagging in J.W.'s chest. He thought and thought, finally realizing it was the local

CO's counsel, Schwartz-something, that Vock was going to let him finish the course then stick it to him.

The company of Rangers, each divested of just-purchased cartons of Mars Bars, Snickers, and Three Musketeers, was forced to stand at attention with eyes open as the contraband was tossed into Big John's pen. With the ground still trembling, the troops were issued half-pilfered boxes of C-rations then ordered to move into the swamp. A student-commander was appointed to engineer a massive assault on the remaining aggressors, those alive after months of unrelenting attack by the Rangers of Class 68-B.

A dozen cadres-partisans gathered at the bleachers. Dressed as Mexican peasants with sombreros and turn of the century muskets, they provided intelligence in Spanish.

Gillette translated. "Rangers, the partisans say the damn Communists are serious about taking over the country. None of us are going home until we stop them. Our partisan friends here are worried. They say the enemy is dug in, heavily fortified with automatic weapons and artillery, and prepared to fight to the death."

J.W. nudged Gillette, "Ask them if they can really do that."

Gillette shook his head, whispering, "It's just a euphemism."

J.W. turned to Branch and asked, "What does euphemism mean?"

"It means they're going to be screwing with you until the end, that's what."

All three platoons set out as one force on the final mission, tramping into the deep swamp then crossing now-familiar cypress stands. J.W. had named some of the trees after former girlfriends. He laughed that the trees, by and large, had treated him better than had the women. He had also grown fonder of the flora than most of the ladies he'd dated.

The company drifted through the slough for hours. They climbed out near dusk and took trails in the savannah they had beaten many times. They passed huge, individual blades of grass that J.W. recognized, freaks of nature taller and tougher than the rest. He smiled nervously, realizing they would still be there long after he was a half-a-world away.

Second Platoon encountered harassing small arms fire soon after departing the water, so Branch, Bearchild, and J.W. were sent to guard the rear. It wasn't five minutes before they captured two of the snipers. The prisoners were uncooperative, and J.W. argued vehemently for tying them to a tree and leaving them for all eternity. Branch, however, insisted on taking them as bargaining chips, and the three wound up putting ropes made of vines around their necks and dragging them when they refused to walk.

As the hostages screamed for help, J.W. tried to gag them, and one bit J.W.'s index finger. A flap of skin pulled loose, and J.W. shrieked he could see his bones and tendons. J.W. let go to tend to his wounds, and the prisoner sped into the swamp. The other prisoner's eyes brightened. He laughed with a heavy Eastern European accent steeped in derision, "Now, you die, Student Ranger, pig!" He looked up in the sky and smiled.

"Fuck you, asshole," J.W. sneered, but an alien vibration filtered threateningly through the triple canopy vegetation. Air surged over them in silent, staccato waves. J.W. feared it was Cowsen, Esposito, Poliak, and Vock closing in on him. Instead, there followed a horrifying crash, and then a clap of screeching thunder rolled over them so brutally, no one was left standing.

J.W. thought at first it was one of Big John's cousins, but the whirling jungle air became a tornado and knocked Branch to the ground a second time. A trickle of blood dripped from his left ear as a dark shadow raced over them. A flight of fighter jets had crisscrossed just feet above the tree tops, the blast from the Phantoms' afterburners driving the ground troops into the mud. J.W. jammed his hands over his ears.

The three Rangers dove underwater as the next strafing run began, though it did little to mute the bombers' unbearable din.

Branch lifted his head out of the water for a breath and spied his prisoner loping toward the edge of the swamp.

Despite the third swoop of the Phantoms, Branch raced after the POW and tackled him, lashing the man's bootlaces together. As he dragged the aggressor back to their position, the air resettled into a sultry stillness. As fast as the jets had come on station, they disappeared.

Branch thought aloud, "That first guy's back to his unit by now. He'll bring 'em here in a minute. We gotta move. What'a we gonna do with this one?"

J.W. hollered, "What'd you say?"

"I said, 'What should we do with this butthole?'"

J.W.'s eyes narrowed. "Kill his ass, and there's one less to keep me from goin' home. Let's try that stranglehold on him."

Bearchild looked up and grunted. "Let 'em go. You got enough trouble as it is."

Branch agreed quickly, and the three Rangers watched as the aggressor walked off haughtily, pumping his middle finger in the air over his head as he vanished beyond a web of Spanish moss. The three broke into a sprint, not stopping until they reached the banks of the Green River. Without dipping their canteens into the fittingly named waterway, they waded across, trudging into the dunes past old objectives they had fought so hard to take, patches of worthless land now devoid of life, spirit, and learning—a football stadium on Monday morning.

Weathersby chuckled sadly at the mounds of sand and empty C-ration cans, wondering what it had all meant. Was it just one of his ambulatory dreams that good men had trudged for two months and then failed so close to the end? Would that be the end of a military career, and maybe a stain for life? Had that misery any consequence for those still there? What had it all meant?

The sun dropped beyond the Gulf, but that night, no moon took its place. The sky turned dark with clouds as black as J.W.'s heart. Then rain came—his first monsoon. He had never imagined water pouring from the sky with such tyranny, or that the dunes could be drowned so abruptly. Ponchos, what was left of them, were as thin as rice paper, and rain poured through the large rips from the night their wet weather gear had been dropped as parachutes.

The precipitation was a scheme concocted by the cadres, J.W. suggested. "It ain't happenstance, I'm here to tell ya. You notice it didn't rain the nights we almost died of thirst, did it?"

Bearchild agreed. "Water can be culled from the heavens if the right person utters the appropriate invocation."

"You're nuts. Both of ya'. Listen to me," Branch demanded. "It was those fighter jets, the Phantoms. What do you think they did when they left here? You see 'em flyin' straight up? They seeded the clouds. This shit be experimental rain, man. You're poisoned for life. Army don't care. It's the same shit they pulled on those black soldiers when they gave 'em a dose on purpose just to see what happens. Wasn't that long ago. Army's the Army."

J.W. shook his head. "A dose? A dose of what? What the *hell* you talkin' 'bout?"

Branch snarled, "A dose of syphilis, man. It was the same shit they played on us back at Harmony Church. You white guys. Everything's just hunky dory, isn't it? Maybe that's why we call you honkey."

J.W. shook his head. "You mean they gave us syphilis at Fort Benning? You're nuts."

"No, fool, I'm tellin' you, army don't care about the health of its soldiers. They gave black troops the syph twenty years ago. Maybe you never heard about it."

"No, I never did. It's nuts."

"It isn't nuts. Army infected a bunch of black soldiers to see if some drug worked. I'm tellin' the truth."

J.W. looked to Bearchild. He nodded and murmured, "It's true."

Branch went on. "And now they're using radioactivity on us. You remember the blue and the green light around the wires when it rained? That's radioactive shit, I'm tellin' ya'."

J.W. hunkered next to Branch. "Okay, FTA—fuck the army, I can deal with that, but it was an electric field at Benning, not plutonium."

Bearchild shook his head. "And both of you, calm down. It's rain—just rain. It rains in Florida. We need to find the platoon."

When Bearchild and J.W. started off, Branch folded his arms and refused to budge, his face frozen in a sneer. Weathersby glanced over his shoulder. "He looks like Mussolini. Hey, El Duce, let's go."

The three moved through rain that beat with such passion, they could not hear each other's boot steps. Instead, they walked shoulder to shoulder.

J.W. voice wavered. "This is the darkest night of my life."

Bearchild sighed, "I hope for all of us, it is."

They caught up with the company just as the lead elements were funneling together, a dark hillock to the left, the Green River to the right. First Platoon pushed through the bottleneck without incident, but as Second Platoon entered the conduit, a series of shattering explosions erupted, hurling them into the mud. Though the flood of light from bursting plastique obliterated night vision, their eyes were still sensitive to movement, and J.W. was able to make out, along with the starburst pattern that had burned into his retina, the stir of shadowy forms descending from the flanks. One rushed J.W., grabbed his fatigues, and slammed him to the ground.

The attacker grunted in a senseless, foreign babble, while another gibberish-hollering man jumped on J.W.'s back, threw a rough-woven nylon bag over his head, and tied it tightly around his neck. Next, J.W.'s hands were bound behind his back with fishing line cinched so harshly, his fingers became swollen then numb.

J.W. kicked and screamed, but with each flail of his legs, he caught a rifle butt in the kidneys. When he decided to remain still, he was pulled to his feet, shoved along, and after a mile, thrown into a thicket. He struck his head on a low-hanging branch that knocked him to the ground. He struggled to his feet, stepped into a hole, and went down again. The bag over his head muffled his screamed refusal to go on, and he drew on the tactics used so successfully by the enemy soldiers he had captured few hours before. He dragged his feet, but these masters let him drop to the sand while they pummeled him.

J.W. envisioned Kenyon being kicked to death, remembering how little impact the loss had had on the world. He struggled to his feet.

His captors pushed on, dragging and shoving him, halting in an open area where J.W. sensed a cooler and drier earth. Cs cooked over a crackling fire, releasing a magnificent aroma, but the cues bidding warmth and comfort were subsumed far beneath his panic.

J.W. was jostled into an earthen pit so narrow and short, he could not lie down. When a woven bamboo cover was dropped over the hole and an aggressor stepped on it, he was forced to stand doubled over, making his abdomen burn with the fatigue of a thousand sit-ups. When the pain in his neck radiated into his arm, he turned his face sideways, but the sharp plait of the cover rubbed his cheek raw. He was left to spit out the blood that trickled into his mouth.

The foot of water in the bottom of the hole soaked his mottled leather boots, turning his feet into soft clubs. He stood flexed at the waist, the pain building until he screamed he was going to kill them. A captor reached through a rent in the bamboo cover, lifted the bag, and shoved a sock deep into J.W.'s mouth. He gagged up the remnants of a long-forgotten meal. There was no choice but to swallow and hope it didn't trickle into his lungs.

Minutes later, he was dragged out of the hole and lifted by what he could tell were six men and dropped feet first into freezing water. His hands were left tied behind him, and his ankles remained bound. One of the enemy reached under the bag and pulled the sock free.

"What is your unit, prisoner?"

"My name is J.W. Weathersby. I have no rank. My serial number is OF116336, and you can kiss my fuckin' ass."

J.W. heard the slap through the bag before he felt it. "What do you mean you don't have any rank? Are you a soldier?"

"J.W. Weathersby. No rank. OF116336. I told you to kiss my mother fuckin' ass!"

J.W.'s head was shoved into the water and held there for a few seconds. His legs had nowhere to go, and he realized he was in a barrel. He coughed and spit at them, "Fuck you and fuck your mother!" the bravado serving to numb his fear, and, oddly, the pain. His head was thrust back, and a wooden cover slammed down on the barrel. He felt the shock waves as it was hammered shut.

Though there was a foot between the top of the water and the lid, some of the putrid fluid leaked up the bag into his mouth and trickled down his throat. They had taken it from the slough, but how had they had made it so cold?

With his head sideways, he sucked in squirts of air through clenched teeth. The lid did not budge as he pushed with his head, and on a second attempt, he drew in a deep breath, drawing along with it a gulp of stinking water. He coughed so hard, he became lightheaded. In the haze, he accepted he was about to die at the hands of ignorant men because he had worn a tiny swath of green instead of gold at a wedding a decade before. His life had been nothing, and already it was over.

He waited for the gentle death by drowning he'd read about, but when nothing happened, he became furious, wanting the travesty of his demise and that of Kenyon's and Smith's to be recorded somewhere. His rage deepened. No, he would fashion the chronicle with his own words, goddamnit. His mind exploded, and he began to cry. He fought like a madman, as frantically as he had to escape the gas chamber. When he got nowhere, he thought about that cocksucker Vock, and J.W. understood this was the final piece of the fanatical captain's plot to destroy him.

After another massive breath, he pushed even harder, but still the lid did not give way. Again, he turned his head to suck in what air floated between the top of the barrel and the water, but the anger ebbed into the loss of all sensation—time became distant and slow. His lunacy of attempted escape had used up the oxygen in the gap. The concentrated carbon dioxide narcotized him, and in what was left of his life, J.W. heard banging on the side of the barrel. The top was pried off.

J.W. admitted to the interrogators he was a lieutenant, justifying himself by reasoning, after all, it was his real rank, and that was one of the four bits of information he was permitted to divulge. Then he conceded that he was in Ranger School, and one of his captives offered a damp cigarette.

Next, they pulled the cigarette away and demanded to know if the African soldier they had captured was a slave, for he had died under their interrogation, and they wanted to know where to ship the remains. J.W. exploded, "No asshole, he was a scholar. Got a master's degree from Penn in sociology. You fuckers killed him?"

A hand jammed his head back underwater, the lid came down, and one of the enemy carped, "So this is the one, huh? The one we know about, the one whose wife Krista is fuckin' the guy downstairs? Get the bag back on this ugly fucker's head and move him the hell outta here. We got dozens to go."

Crushed in the cell, J.W. closed his eyes and cursed himself for having gone beyond name, rank, serial number, and date of birth, all for a cigarette from which he'd managed one short drag. He tried to imagine Branch gone, but his musings were cut short as he remembered the bastards had blurted Krista's name.

His mind roiled, and he slowly came to accept it was, yet again, but a dream. He settled a bit until an aggressor came to his cage and banged a metal garbage can lid against the bamboo roof. J.W. tried to raise his bound hands over his ears, but he could not squeeze his arms past his shoulders. The man reached into the cage, pulled the bag free, then stood and unbuttoned his fly in front of J.W. He moved a few steps out of Weathersby's line of sight. A stream of water dribbled onto J.W.'s head.

The captor walked to the front again and waited for his hostage to look up before buttoning his fly. The man spoke in a heavy accent. "I have letter from your woman. We capture it. You worthless, weak soldiers. Now I read letter. Even your woman piss on you.

"Dear Baldini," the enemy soldier laughed moronically, "I just wanted to tell you how much I missed you at first, and how hard I tried to stay good. Then the fellow downstairs, you remember Gordon, he helped me get the Volvo started, and things just happened after that. Now you're due home in a few days, and I don't know what to do. I had needs too, you know. Now I can't stop. I'm really sorry, but I had to tell you. Signed, warmly, Krista."

J.W.'s spirit sagged into a black pool of despair. He had never tasted such desolation; he had never dreamed a heart could weigh with such agony. Even Yonah Mountain was bliss compared to what they had done to him here. It was clear the guard could not have made up such a letter. No one in Ranger School knew his wife called him Baldini, a reference to his thinning hair—no one, except Branch. J.W. had never told anyone about the problem starting the Volvo—but Branch. So that was it. Even Branch had collaborated with the enemy.

The life boiled out of him, and he became numb. He castigated himself for having conceded he was a Ranger, a detail that made it clear all of the prisoners were Rangers, for commandos didn't travel alone. These men were cunning. What would the next

Ranger give away? Would they turn around and use it tomorrow, or the next day, or the next year? Would there be another letter, maybe from his parents announcing they were getting divorced because their only son was a disgrace for having given in and become a quisling?

Had J.W. not hurt so badly, he would have laughed at his frailty. Instead, the pain in his chest and viscera worsened, and he shouted, though this time no one stood over him banging metal or urinating on him. When the pain eased for a moment, he thought of the letter and realized he had never told Branch the man's name downstairs, or even that there was a downstairs. It was real.

J.W.'s head drooped further, and the sadness consumed him more profoundly than any emotion he had ever experienced. It only worsened as the groans of other Rangers being drowned filtered through the POW camp. He heard a lot of "Fuck yous" and "Suck my dicks," followed by hammering and voices finally grunting, "He's not a slave, asshole."

One by one the voices came and went until there was a, "Kiss my dick, fool. I ain't no slave, mutha fucka." After the pounding and an even longer wait, punctuated with deep fits of coughing and gagging and another stream of profanity, J.W. caught a final, feeble, "I told you, kiss my dick, honkey," then silence.

The nailing started again and anger welled in J.W.'s chest. It was the same rage that had seized him on Yonah Mountain. The mortifying voices boomed, though it was he who was roaring this time. "If I fuckin' die tonight, it will be at the moment of my choosing, not yours, mother fuckers."

He squatted deeply and exploded upward, flying out of the pit, snapping the bamboo roof off the cage easily. His vision caught an edge—it had been fastened by only a single filament of fishing line. He crawled from the hole toward the sounds at the interrogation chamber. He pulled bound hands around his legs to his front, tore at the thin string with his teeth, then looked up to see two Green Berets standing by a wooden barrel laughing. J.W. swung his head back and forth, searching for the company of troops that had captured and tortured the Ranger prisoners. Aside from the pair at the barrel, he spied only one man standing over a cage, pouring water from a canteen on the head of his captive. The camp was otherwise empty.

J.W. picked a hefty stick from the earth and moved from palmetto to palmetto, coming up behind the closer of the two enemy commandos. He watched as they dropped the lid on the barrel, and he waited for the hammer and nails, but one aggressor just picked up a small rock and tapped perfunctorily on the edges.

J.W. crawled forward, jumped to his feet, raised the stick, and whacked it into the back of the aggressor's head so hard the wood cracked. The solder's beret flew off, and the man fell to his knees. J.W. raised his weapon to strike the second man, but a dark figure burst through the wooden barrel-top like a jack-in-the-box.

It was Branch. His loosely bound arms shot out of the water and grabbed the standing guard's throat, seeking with all his might to crush out the man's life. Enough water splashed out of the barrel onto the kneeling Green Beret's head to shock him back to his feet in a frenzy of swinging and cursing.

The ground surrounding the three-quarters-buried barrel had become a mud pit, leaving J.W., Branch, and the two Green Berets flailing in the mire. The third soldier arrived swinging frantically at the Rangers, but he bashed one of his pals in the melee. Another of the GIs managed to get a forearm around J.W.'s neck. He squeezed until Weathersby's face hued into a deep blue.

"Stop moving, Ranger, or I'll fuckin' strangle you all the fuckin' way. Got that, asshole?"

Out of the corner of his eye, J.W. saw Branch wallop his adversary in the head, jump behind him, and jam an arm around his neck. With his last breath, J.W. groaned, "Crush the mother fucker's throat, man. I don't give a fuck what happens to me."

The Green Beret holding J.W. screamed, "At ease, both of you! Calm down, or your fuckin' Ranger buddy here dies."

With the "Fuck this's" and "Fuck thats" booming through the prison camp, a chorus of hooting and ranting rose from the holes and cages.

"Now look what you assholes have done," the troop squeezing J.W.'s throat cried. "This is part of the last problem, to be captured. Every Ranger has to go through it; otherwise, you're not a Ranger. Jesus, look what you've done, you fuckin' asshole."

The five soldiers lay breathing fiercely, each slowly relinquishing his prey, scrutinizing every infinitesimal move to insure the process of disengagement proceeded bilaterally.

When all arms and legs were free, J.W.'s Green Beret lit a smoke. In the match light, Weathersby saw brown and green in the man's eyes. "Release the rest of your buddies and get the fuck outta here."

As J.W. moved toward the cages, one of the soldiers ran up and grabbed his arm tightly. "Ranger, you know who that is, the guy you been fuckin' with the whole time?"

"Yeah, an asshole who won't leave me alone."

"That's not the way he tells it. First Infantry Division, Viet Nam. Operation Shenandoah II. Up for the Medal of Honor. You have any, *any* fuckin' idea what that means? Do you know what you almost just did?"

J.W. breathed out a long sigh. "You tell me you killed my Ranger buddy, and I shouldn't fight back? Just let you murder us? Already did it back at Harmony Church. Yeah, I'll remember that in a week when *I'm* in the Nam. And anyway, he should be on recruiting duty, or talkin' to Congress or somethin', not getting' his ass beat every other day in the swamps. I thought the army protected those guys. What the hell's he doin' out here?"

"Tryin' to train you jerks, so he doesn't have to do it again."

⊨⊨

J.W. and Branch went from underground cage to underground cage. None of the bamboo tops had been secured to the cells by anything more than a single filament of 5-pound-test nylon. Morelotto was silent when J.W. cracked the roof off his hole. He held his right arm defensively, close to his body, but let J.W. push up the sleeve of his fatigue jacket, the thin, worn material more gauze by now than cotton cloth.

His forearm was curved in an ellipse, like the crescent moon that peeked from behind thunderclouds. A tiny point of jagged bone poked through the skin.

Gillette examined it as well. "Compound fracture of the radius. Shit."

J.W. frowned, "That must hurt like hell. We need to get you back to Field Seven."

But Morelotto grunted through gritted teeth, "After the attack. I'm going to have to walk out anyway. They can't get a meat wagon or a chopper in here, and I'm too fat to be carried." As J.W. turned to Gillette for counsel, Morelotto added, "Hey, Weathersby, thanks, man."

J.W. searched futilely for Bearchild as the imprisoned Rangers gravitated together and fell into formation. The Ranger who had been in command when the company was captured took control and grouped his men into a circle to plan the final attack. J.W. waited for the dissension when a forced march was proposed to make up for the time lost in captivity, but there was only silence. Second Platoon of Ranger Class 68-B nodded in agreement, unanimously, not to allow their student commanders to be blamed for leading them into an ambush.

As they gathered their gear to move out, a Green Beret handed J.W. an envelope. Wet and faded, it was a letter he had never been given.

*Dear Baldini,*

*Having trouble getting the G-D Volvo started, so Gordon, the little guy downstairs, you remember, the nervous airline pilot, he looked at it and found the wrong something or other. It's great now! I love you. Can't wait to see you in a few days!*

*Will write again tomorrow. I hope you've gotten my letters. I've sent one every day, even the day you left. You remember Herman Hancock, the navy guy who was trained with you in flight school? His wife, Lynda, told me you aren't allowed to write during Ranger School. I want to hear all the stories. Bet you got a few.*

*Love, love, love,*
*Krista."*

J.W. reread it a few times to be sure and saw it had been postmarked only two days before.

The company moved quietly through the jungle, though almost at a jog. Branch and J.W. hung behind, taking turns carrying Morelotto's gear, dragging over deadfall and prickers, always keeping an eye and an ear out for Bearchild. In the savanna they traveled worn, but dangerous, paths. Even with Morelotto in tow, if that pace could be maintained, they would hit the final objective in four or five hours.

"Gentlemen," the company commander counseled during the final briefing, "we know the terrain, we know the enemy, and we know who we are. No more Rangers gonna be droppin' out. We made it. Let's kick their ass!"

J.W. called to Gillette, "It's funny, man, I'm not so hungry, not so tired."

Gillette laughed, "That's what I was trying to get across back at Harmony Church, the lesson about the Vietnamese at Dien Bien Phu. Remember, when everyone pissed and moaned? And how 'bout that old sergeant major?"

When they stopped for breaks, men stood and oriented themselves on their maps. No one lit up. There was silence during movement; even the Ranger Bird had been silenced. Branch smirked, "Sucker's gone extinct."

They waded into the jumping-off point for the final assault in three-and-a-half hours, a class record. Branch and J.W. hid Morelotto in a safe cove, placing the shattered arm in a splint fashioned of sticks and the remnants of J.W.'s tee shirt, then covered him with palm fronds. Branch had two cigarettes left. He gave them to Morelotto, who shook his head. "No, didn't earn 'em." Branch gave him a tap on the shoulder, then dropped the smokes and matches next to his good arm.

Bearchild had waited for the platoon high in a cypress, lounging on a limb like a serpent. A rumor made the rounds that he had escaped the pincer movement by wading into the river and breathing through a hollow bamboo stalk, a trick some of the Sioux had used to escape the U.S. Cavalry at Wounded Knee. When the coast was clear, he climbed out of the river and followed the bound Rangers, plotting to attack the aggressors from their rear and free his compatriots. As he spotted more and more of the enemy filtering through the woods, though, he accepted he was outnumbered. And anyway, he reasoned, all of them would be freed to carry out the final mission in a couple of hours, for if the date on his hidden calendar was correct, graduation was scheduled for the next morning.

So Bearchild spent the hours scouting enemy positions ahead of the company, gathering crucial intelligence: numbers of troops, their dug in positions, and where the cadres were hanging explosive charges in the trees. He debriefed the final Ranger commander, explaining that he had hidden in the bushes outside the cadres' camp, listening as they drank coffee and planned to man the final attack on the cheap, sending most of the aggressor GIs back to Field Seven to prepare for graduation.

He went to J.W. and whispered, "Vock was there. I heard him say, 'I'm gonna get that son of a bitch at the final formation. Watch me now.' The rest of the cadres laughed." J.W.'s body stiffened. "Sorry, man, but you needed to know."

And Bearchild had gathered more than gossip. One of the sergeants had left a tactical radio near the perimeter of the aggressor camp. Bearchild liberated it. Taped to the side of the PRC 25 transmitter was a directory of frequencies for the aggressor units. Bearchild tore off the list, put it in his pocket, then used the radio to guide the Ranger company to his position. As his colleagues approached, he left their frequency for a moment, and while beating his chest with a fist to the rhythm of a helicopter rotor, called the enemy camp and told them he was the supply pilot and needed directions to their position. The enemy radio operator provided not only coordinates but troop strength, ammo, food requirements, and in the end, when Bearchild radioed he couldn't find the camp, the cadres popped a willy peter grenade, the white phosphorous lighting their position like the Fourth of July.

Bearchild flipped frequencies and directed the Ranger company through the jungle to their final objective, an open savanna less than half-a-mile away. Though the cadres had placed listening posts, every time one of the forward scouts radioed in that Ranger units were inbound, Bearchild intercepted the message and howled like a wolf into the microphone, obliterating the transmission.

As the Rangers broke into small tactical units for the final attack, they found the aggressors lounging peacefully in hammocks. The enemy faced north, for no company of Rangers had the stamina at that point to sweep wide in an arc and hit them from behind. There were, the cadres laughed, certain inviolate truths—Rangers invariably trod the easiest path toward mock battle.

Class 68-B, however, approached soundlessly from the south and swept in like the divine wind, perfectly silent until they looked intently into the whites, and greens, and browns, of their enemies' eyes. As they dashed through the encampment in victory, the Rangers took prisoners and confiscated weapons, binding the hands and feet of the captured, concentrating them in a circle at the middle of the compound. The Rangers then set about un-earthing the spoils of war.

J.W.'s attention was drawn to a tree with a patch of gnarled bark whose lines were not exactly parallel to that of the rest of trunk. On the ground were telltale bits of fresh wood, and J.W., assuming from his lessons in elementary school that termites didn't eat live trees, touched the bark with a quavering finger. The scrap of wood fell away, revealing a small niche into which he shined his flash-light. Twigs and leaves were packed as the next ruse, and he picked them away to stare at the business end of an OD can. He pried at it greedily with his bayonet until it popped out and tumbled to the ground. He invested in the process of squatting and lift-ing the potential trophy and moved quickly through the balance of the algorithm: the can had mass, the top end was untouched, the bottom was chaste, and the walls of the precious cylinder were pristine. A feeling of comfort washed over him, a bowl of macaroni and cheese when you were sick.

He had captured an olive-green, black-lettered tin of Ham and Muthas. His imagination flashed to the sigh of air rushing in when the P-38 pierced the virgin lid, and then to the perfume of several-year-old wisps of fatty ham and crunchy beans wafting toward his gaunt face. What should have been ecstasy, though, degenerated into uneasiness. He searched for Branch, and seeing him stagger-ing mindless circles in the slough, waded over, the can held aloft. Branch's eyes reddened.

The aggressors, bound in their ring of shame, snorted at the Rangers scraping away at the ground and trees. "You jerks'll be payin' real soon like. You rough us up, you get your asses beat when the cadres get here. There's a helicopter on the way with fresh lane graders. You assholes are fucked."

The vehemence of the threats implied, Gillette suggested, that there was additional hidden food, and several Rangers were dis-patched to unearth it. A carton of C-rations, carelessly hidden under leaves in a punji pit, was carried back intact. Branch, energized by the recent calories, divvied up the swag with great decorum.

The aggressors slung epithets at the Rangers, the black troops yelling out to Branch, "Hey, looky looky at that shit—slave be back doin' for da honkey. Hey, Uncle Tom, man." They threatened Bearchild, "Hey you fuckin' slant eye gook, bring me a meal, or I'll put your ass back on the boat to Viet Nam." They kicked dirt at the Rangers, and one man lifted great hockers of spit, looping them into the air like mortar rounds. The cadets laughed and drew an X in the sand where the most energetic of the wet bombs touched down. The Rangers took baby steps beyond the target and flipped off the bombardier.

It was time to open cans and suck down congealed meals, all of which were gone in thirty seconds. The evidence was tossed far into the swamp, mouths were rinsed, and listening posts sent out to warn when the ranking cadres approached. But most of the Rangers sat about, just out of spit range, smoking and talking, staunchly believing the course had come to an end. They ignored the ranting of their prisoners, for school was out. They were the officers again, each of them worthy of a Ranger Tab.

Branch and J.W. went back to the cove and fed Morelotto Spaghetti and Meatballs—his nectar—then guided him to the victory site. He sat by himself, silently holding his arm, staring into nothing.

Bearchild watched him for a moment and offered aloud, "Looks like a Ranger to me."

A moment later a file of embarrassed cadres crashed through the jungle, Vock in the lead. He snapped, "On your feet. Fall in."

The released prisoners met with the cadres and pointed to several Rangers. They shoved off an instant after Vock hissed, "You men got the rest of the night off. I'll take care of business from here."

Morelotto was ordered to stand to the side but wasn't tormented. Cowsen appeared out of the vegetation, rubbing his chin while the cadres picked out individuals to reprimand. At one point, only three Rangers were on their feet, the others in the pushup position, working off gigs for six-dozen unrelated crimes. J.W. was awarded twenty-six for not wearing glasses, and Gillette twenty-six for wearing them. Branch caught eleven for a stubble of kinky hair; Wardally tallied twenty-six for wearing the patch of the Grenadian Army, and Bearchild just ten for straight hair. Each man screamed cadence at a different rate, nary a man able to stifle his laughter.

Cowsen stood before them and shouted that the attack was the sorriest performance he had ever witnessed from a Ranger class, and that they would do it over, and then over again, until they got it right. "I don't give a hoot if that means stayin' an extra eight weeks, Rangers. I got plenty of time. I'm not goin' anywhere. And I'm not done with you for the way y'all treated your M-14s. They's muddy and filthy. Gentlemen, that weapon is your lifeline. You clean it first, before you take a drink from your canteen. What a sorry company of so-called Rangers."

Eventually, Cowsen about-faced and ordered, "Left, face. Y'all got a long march back to main camp to resupply and do the whole cotton pickin' mission over."

Wardally wound up at the head of the column. He passed his machine gun to J.W. at the rear; ammo went to the next to the last man, Branch, and the radio to Gillette. The heavy gear would be rotated during the march. They reached into their packs, removing the flotsam and jetsam of their lives. Tossed into the culvert by the side of the dirt road went second helpings of pilfered Cs, water purification tablets, smokes, and sugar. J.W. fished out the half-tin of Beef Stew he had obliged himself to save. He offered it to Branch, who made an ugly face. The gravy came out in a shiny, long, gluey, brown strip as the can flew into the brush.

Vock walked into formation, placing his lips a quarter-of-an-inch from J.W.'s. "Did I see you destroying government property, soldier?"

"No, sir!"

"You're lying, soldier! Step out of formation. Sergeant Cowsen, did you see this man throwing government property into that ditch?"

Cowsen hesitated. His lungs filled slowly, wasting seconds, gazing up into the jungle canopy. When he was out of time, his lips parted and the words began, "Sir, in regard to your question, I..."

He stopped as a commotion came from the road. An Air Force sedan, flanked by military police jeeps, rolled up to the decrepit formation. Flying from the main car's front fender was a small red flag embroidered with three white stars. Vock spun around and came to rigid attention, as did the cadres. Vock saluted the car, then, as ranking officer of the Ranger contingent, ran up and reported.

The base commandant bristled from the car and opened the door for a civilian in a Brooks Brothers suit. "Captain, I want you to meet the Secretary of the Air Force."

Vock remained at stiff attention as he shook the civilian's hand. Morelotto, who was slumped at the back of the formation, was waved over and tried to salute the commandant, but his arm collapsed to his side. The civilian rolled up Morelotto's sleeve and gasped. So did Vock. There was muffled talking, and Morelotto pointed animatedly with his good arm toward Weathersby, Branch, and Bearchild. The corners of Vock's mouth turned down as the civilian summoned the three ragged officers.

Morelotto stood at near-attention next to his father and spoke quietly. "I wouldn't be standing here now if it wasn't for these Rangers." He paused and looked up into the grey sky. "Dad, I probably would have died out there. They're real friends, and they're real Rangers. Army's lucky to have them."

Cowsen, his face as calm as J.W. had ever seen it, swaggered to the front of the company, sounded the order to march, then upped it to the double-time. Second Platoon, which found itself at the rear of the formation, ran a quarter of a mile and passed the big guns one man farther up the file. At that moment Third Platoon, in the lead, rounded a blind curve, screamed, and broke into a mad sprint. Assuming they had been hit with chemical weapons, someone screamed, "GAS!" and the men pulled masks from their packs.

As the next platoon rounded the curve and broke into a sprint, howling and screaming, J.W. bitched, though no one could hear him through the mask. "They have to squeeze the last goddamn ounce of blood outta us? I will quit this shit before I execute another fuckin' attack."

A hundred paces later, it was Second Platoon's turn to round the poisoned blind curve. J.W. adjusted his gas mask. But there was no gas, or machine gun fire, or green snakes. One hundred meters further along the road were three cattle trucks, the first two swarming with delirious Rangers. Second Platoon ran to the third vehicle, clambered aboard, and dropped onto the wooden benches.

Cowsen and the corps of cadres followed at a relaxed trot. There was complete silence as the first sergeant stood in the middle of the road, all eyes glued to his STRAC figure. "Y'all still look like fried doggone chicken!" But Cowsen snapped to attention and saluted each truck in turn. The bedraggled Rangers came to attention, returned his salute, then screamed as they tossed their hats into the air.

The trucks drove slowly through the mud and humidity. J.W. was quiet the miles back to Field Seven. As the trucks pulled up to the barracks, he spoke to Branch. "What just happened?"

Branch thought a moment. "I don't know, man, but I think it was important."

Bearchild nodded.

The company was dismissed to prepare for graduation rehearsal. That boiled down to brushing their teeth. They marched at leisure-time to the runway on Field Seven, bitching that the ride had softened them, allowing that they were embarrassed for their churlish behavior over the months.

Bearchild observed, "You're still complaining."

Wardally threw his arms up. "You Americans are never satisfied. In the Islands, we don't suffer guilt when we've earned our rest."

Preparation for the final ceremony consisted of a sloppy jog around the four-mile runway at Field Seven. It was done to the cadence of relentless grousing.

After the run, halting at the reviewing stand, they stood dripping in the noonday sun, trying to remember to salute with the right hand and accept the diploma with the left. J.W. asked why he, a southpaw, couldn't salute with the left and take the diploma with his right. The cadres didn't laugh, but J.W. had two years before when his grandma had asked the same question at his commissioning ceremony.

At the barracks, the troops lined up at the dumpster to toss away their ruined boots, a ceremony carried out with great dignity. The men offered their farewells and thanks. J.W., however, had worn his dress boots in the field for the past two weeks, having long since buried his original footwear in the dunes. Then they changed into the one clean pair of fatigues each had dragged through the three phases, the uniform they had been directed to keep wrapped in paper bags, dress right dress, at the ready, as if graduation could have been declared at any moment.

There was no hot water for showers in the barracks because the fuse on the water heater had blown on the first day, and no one had reported it. After the freezing water, they double-timed back to the "parade" ground and ran the entire four miles again, down one side of the two-mile runway, across it at the end, then back up the other side, finishing, soaked with sweat, at the reviewing stand. This time the benches were packed with onlookers, men and women, most in civvies. Cadre representatives from the three phases were there with wives and girlfriends.

Vock sat alone. J.W. reassured himself that in a few minutes the US Army could no longer hurt him. There were no punishment schools left with which to torture him—he'd be in Viet Nam by the end of the month, no matter what he did or did not do. He grinned, planning his final act at Ranger School: accept his graduation certificate, glare at Vock for a moment, then laugh in his impotent face.

When the ritual began, a lieutenant general, the post commander from Fort Benning, took the podium. He told the gathering that he had flown all the way down just for this particular graduation—it was that important. He spun several yarns, the most captivating, a vignette about a Ranger cadet he had met sometime in the distant past, a lost soul who hadn't yet learned the potential of the phrase, "DRIVE ON!" He mused aloud about that man. "I wonder if he ever discovered the power in those words."

Cowsen nodded ever so slightly to the general as J.W. accepted his diploma. When J.W. shook hands with the commandant, the man leaned forward and growled, "*Now* you're ready to lead men in Viet Nam."

＝＋＋＝

Next to Big John's pen sizzled a barbecue with more beer piled in ice-filled garbage cans than J.W. had ever seen in one place in his life, even at his fraternity. He wolfed down so many grilled hot dogs, it was hard to walk. At the billets, J.W. opened the screen door and collapsed on the first cot to the right, not moving until after sunrise the next morning.

Although still digesting yesterday's dinner, J.W. easily wolfed several breakfasts, saving bacon for Big John, who ignored it. By

7 A.M., the company was aboard cattle trucks bound for Fort Benning. They slept away their last few hours together.

⊶⊷

At Harmony Church, the company threw on real uniforms, J.W.'s decorated with the silver wings of an army aviator. Branch and Bearchild were astonished that he really was a pilot. They left for main post together to have Ranger tabs sewn on their uniforms.

When J.W. donned his, he mumbled to Bearchild and Branch, "Doesn't feel any different. Lotta work for a little piece of cloth, man. No big deal, huh?"

Bearchild nodded in agreement. "The melancholia of things accomplished."

Outside the tailor shop, a white-haired sergeant passed. He was wearing the tab. The men saw it when he lifted his arm to snap a salute and bark, "RANGER!"

⊶⊷

As they packed to leave, J.W. was ordered to complete his paperwork for Viet Nam before returning to Maryland. At main post headquarters, J.W. filled out the requisite forms, releasing the military from all responsibility for the flight to Viet Nam, though it wasn't the trip there that worried him.

His veins were punctured by novice phlebotomists to check one last time that the blood type engraved on his dog tags was correct. It wasn't. His tags said "O Positive." The technician claimed J.W. was "O Negative."

J.W. didn't care. Refusing to make a fuss over a plus or minus sign, he asked the private to be a good chap and ignore the problem.

"No, sir. You need to be issued new tags, sir. I can't take responsibility if you get shot in Viet Nam, and they give you the wrong blood, and you go into convulsions or something and drop dead. No, sir."

"Okay, Private, just gimme new tags, and I won't bother anyone anymore."

The technician looked down at J.W.'s clenched fist and withdrew to the back of the clinic. A buck sergeant emerged, and J.W. jerked to attention. Barely suppressing a laugh, the sergeant provided J.W. with the papers for new tags, but explained the next step was to have them signed by J.W.'s commanding officer.

"I don't have a commanding officer. I'm just trying to go home!"

"Who's the CO at the Ranger School, sir?" the sergeant queried, thumbing through the post directory. "Here it is, sir. It's a Captain Richard Vock. Better hurry, sir. It's already 1400, and the dog tag center closes at 1600. It's across post, sir. They ain't openin' again till Tuesday."

"Tuesday? Why not tomorrow? You're open on Saturdays."

"You musta been away for a while, sir. Three-day weekend. Memorial Day, sir."

⚔︎

First Sergeant Cowsen sat behind a desk outside the captain's open door. He stood and saluted. "Yes, Lieutenant, how may I help you, sir?"

"Sergeant Cowsen, I'm trying to get home. Got the wrong blood type on my dog tags. Need the captain to sign these."

"Is that your cab waiting outside with the meter running, sir?"

"Yes, it is. Look, First Sergeant, I don't have much time. They're closing soon for the long weekend."

Cowsen disappeared into the office. J.W. heard hushed mumbling, the dial of a phone, and Cowsen's report. "Sixteen hundred, sir."

Vock appeared in the doorway, and J.W. slid to attention, saluting weakly.

"Come in, Lieutenant." J.W. followed at a distance greater than the length of the captain's arms. Vock signed the form, told J.W. to be at ease, then stood and turned to the window overlooking the pull-up bar.

"Lieutenant, I have a responsibility to insure only the best survive. This school exists to train combat officers who are already good when they get here. You were not. You should have washed out five or six times."

"Why is that, sir?"

"Because you were an also ran. This is not a reform school, Lieutenant. Let me be direct. You should have been gone the first night—taking a ride from the base commandant; then you sneak out of the morning run; then it's a ride in the meat wagon. It goes on and on. You managed to dodge the bullet, the bullets. You made all of us look like fools during the first phase.

"Yes, sir."

"Did you learn something here?"

"Yes, sir."

"What was it?"

J.W. answered weakly, "To drive on, sir."

"Anything else?"

"I'll be frank, sir, if I may."

"Say your piece."

"Okay, sir, the US Army tolerates assaulting and murdering its own. That's what I learned."

Vock turned back to face Weathersby, who took a giant step toward the door and raised his fists. "At ease, Lieutenant. You happen to be right. I fucked up, and I have to live with that. It may even cost me my career. So don't worry, you'll be getting plenty of revenge. On the other hand, that seemed to be a turning point in your stay here, didn't it? They told me you were the last one to start and the first one to crest Yonah. That mean anything to you?"

"Just another hill, sir."

"No, it's not just another hill, Lieutenant. That's the real test. You get to eat steak and potatoes the night before. You sleep on the cattle trucks. So there's no excuses. Yonah's all about what's inside your heart, not your stomach. And you won.

"And, you might be interested to know, charges have been filed against Staff Sergeant Hartack for dereliction of duty in the matter of Ranger Smith and the death of Ranger Kenyon. He was relieved of his duties the day you went to Dahlonega, and he is awaiting general court-martial—in the stockade."

"Sir, he was unfit for duty the instant he issued an illegal order for me to leave my Ranger buddy. Why was he given the chance to murder Specialist Kenyon?"

"We had to get our ducks lined up. But if we have anything to do with it, Hartack won't see the outside of Leavenworth until 1999." Vock looked at his watch and nodded toward the door. "Get going."

J.W. promised the cabby an extra five dollars to deposit him at the Dog Tag Center before 1600 hours. The man smiled, "No sweat," but they were delayed by marching troops, rolling M-60 tanks, and a fire at the BOQ that closed the roads around main post. The cabby shrugged, turned off the engine, and bummed a smoke from J.W.

At the Dog Tag Center, J.W. rapped furiously on the locked door until his knuckles bled. The PFC who unlocked it apologized, "Sorry, sir, we're closed until Tuesday morning, sir, until after the long weekend."

J.W. pleaded that he had but a few days to spend with his wife before shipping out to Viet Nam. A master sergeant came to the door from the back to investigate the increasingly shrill voice. He advised, "If the army wanted you to have a wife, sir, they would have issued you one."

The cabby dropped J.W. back at Harmony Church, marooning him for the weekend in the barracks, three days, a third of the time he had left before shipping out for the war. When he found his personal gear strewn on a chair outside his billet, he charged inside to chew ass, but there were just shavetail lieutenants lying quietly on their beds. The stenciled name on the bunk he had occupied months before read "WILKERSON." Class 68-E had commenced.

He ran to wave down the cabby, but only the car's red dust was visible. Sergeant Cowsen called main post and asked for a jeep to be sent out. J.W. waited the two hours in the parking area, mulling the frigid winter that had just passed and the tropical year that lay ahead.

As the sun began to fade, a loud whistle pierced the quiet of the Ranger camp. A familiar voice screamed, "Fall in," and the next sound was of men knocking on the command shack door. "Harder! They can't hear you in there, Ranger!"

Waking with a start at 3 A.M. the next Wednesday morning, J.W. put one foot in front of the other, avoiding the surrounding cypress trees. He came to one that was curiously white, and he pulled at it. He pried an unopened Pepperidge Farm Angel Food Cake from inside the trunk and dropped to the floor to stuff the whole thing into his mouth with his fingers. He washed it down with half-a-case of Cragmont Orange Soda.

J.W. slept for many days, through Bobby Kennedy's assassination and the riots. Each dream was the same—he was trudging across an endless, icy swamp, the night so dark he feared the cadres had snuffed the moon and the stars.

# CHAPTER 4

# PICKING UP THE PIECES

B efore leaving Fort Benning, Bearchild collected ten dollars from Branch and J.W., added his five, and mailed the cash off to Mr. Sullens, the farmer in the Smokies. Bearchild wrote a letter to the camp commandant asking the man to make sure the Rangers respected the farmers' fields in the Smokies. Branch and Bearchild added their signatures.

In the parking lot on that last day, Bearchild put his arms around J.W., grinning, "I never hugged a man before."

Captain Marlin Bearchild served in Viet Nam, asking the whole tour why he was shooting at people who looked so much like him. He received a letter from an aunt who wrote that his father was up to his old tricks, marching around town jabbering to anyone who would listen that his son was going to be the first United States Senator born and raised on an Indian Reservation.

In 1992, United States Senator Marlin Bearchild retired from politics and created the Steven War Bonnet Memorial Foundation, a watchdog organization that insured the quality of the medical care offered to the indigent—its main tenet: no doctor may ever turn away a patient in need.

Leonard Fricker finished Ranger School with the next cohort, Class 68-C. He served with the First Infantry Division in Viet Nam and remained in the Army for twenty-five years. He retired as a brigadier general. His only child, a daughter, graduated from West Point and flew Black Hawk helicopters in Iraq.

Antonio Morelotto slogged the paddies in 1970, a platoon leader with the Americal Division. Though shot in his bad arm, which he eventually lost, Lieutenant Morelotto carried one of his severely wounded men several miles under withering enemy fire. He was awarded the Distinguished Service Cross by Richard Nixon on August 8th, 1974, one hour before the President went before the American people to announce his resignation. Morelotto went on to law school at Columbia University and later served as a Federal Judge in the 9th Circuit Court of Appeals in San Francisco.

Erskine Gillette was a combat officer in the Central Highlands of Viet Nam. He distinguished himself by leading his besieged platoon over a mountain pass the Viet Cong never dreamed an American unit would or could traverse. He returned to Harvard to earn a Ph.D. in philosophy, though secured tenure in the Department of Asian History there. He died of leukemia at age forty-two.

Big John the Alligator passed away of natural causes in 2001. He was originally captured in 1954, just half-a-mile from Field Seven. As lore had it, he lost his eye in a struggle with Rangers who had happened into his domain on the Green River. Truth be told, John was a she, and the eye was lost when the grass was being cut in her pen with a weed eater. Her remains were interred in a corner of a field near the old pen. A small headstone marks the spot.

J.W. saw Lawrence Ellsworth Branch once again after Ranger School. It was in Viet Nam. J.W.'s year of combat over, he was on line at the air base just outside Saigon, preparing to board the Freedom Bird home. Branch was on line with the FNG's, Fuckin' New Guys, who had just arrived. It took a minute for them to recognize each other. Branch had gained weight; J.W. had lost even more.

J.W. shouted across the terminal, "Where the hell you been while I been fightin' the war, my man?"

Branch yelled from his barrier, "Graduate school at Penn. Got the PhD, man." Then he waved wildly, smiling so warmly, J.W. had to look away lest his tears taint the voyage home. The two parallel lines were hustled toward their very anti-parallel destinations. J.W.'s last image of Branch was the scar on his forearm from the drifting ember at the little farmhouse near Dahlonega.

J.W. learned in 1995, from the black marble Viet Nam Memorial, that Branch had perished in combat on May 17, 1970. Many years later, J.W. used a computer to track down Branch's mother, Bessie, in Harlem.

"Miss Branch, I've thought about you for many years. How are you?"

"Old. I got the sugar and the congestion. I can't walk too good, specially the steps to Lawrence grave. I take the bus from Harlem to the cemetery, and I stand there at the bottom. What else can I do? You doin' okay?"

"I'm doing well, ma'am. Thank you."

"Well, that's good." She paused. "You gotta job?"

"Yes, ma'am. I'm a doctor." His eyes reddened, and his voice faltered. "I'm sorry, Miss Branch, I gotta go. You did a great job raising him. He taught me a lot."

"Wait a minute!" she whooped. "You say yo name's Weathersby? Ain't you the one who said he could swim forever?"

## CHAPTER 5

# AND SO I FOUND A WAY

In June, 2006, nearly two years after I began pestering the Pentagon to allow me, nearing my sixty-second birthday, to participate in the Ranger Course at Fort Benning, it became clear they would not budge. It did not require a degree in medicine to understand their concern, that a middle-aged man could drop dead on the spot for no reason other than he had been alive for six decades. More importantly, I had to accept that our army was fighting a real war, and my participation would be a distraction to the young men learning to do so. I decided not to push any farther.

The next night, however, while reading over a draft of *At Yonah Mountain*, I came to the part in which the Rangers, back in 1968, believed the cadres were concocting a plan to *march us* from Fort Benning to the mountain camp in Dahlonega. The memory of that night made me laugh, for the thought of them punishing us that harshly, even for those cadres, was beyond the pale. The notion, though, piqued my interest, and I began to wonder if the answer to my craving to relive the challenge of Ranger School might be, indeed, to walk from Fort Benning to Dahlonega. It was midnight, but I couldn't stop myself from Googling the route. Though over two hundred miles, it did not seem all that terrifying. Then, in a fit of absolute madness, I plugged in the entire trek—Dahlonega to

Benning to Field Seven, the jungle facility near Pensacola, Florida. The whole course, on less direct, but safer country roads, was just shy of 500 miles, a distance that would take a bit over three weeks if one ground out twenty or so miles per day. That, too, seemed doable; after all, many older people participate in the Breast Cancer Three-Day, Sixty-Mile March each summer. I was then struck with the notion of finding sponsors and creating a college scholarship fund for the children and grandchildren of Rangers and 11th Armored Cavalry Blackhorse Troopers, the unit with which I had served in Viet Nam.

And so, on 18 Oct 2006, I closed my medical practice and set out on foot, accompanied by an old friend from graduate school, starting at Camp Frank T. Merrill—Merrill's Marauders—the new name of the old mountain training facility near Dahlonega, Georgia. We lugged fifty-five pound packs stuffed with much useless gear and chose not to enlist the support of a "chase car." We were on our own, save for the lessons I had learned nearly forty years before, and, perhaps more importantly, the hospitality of the citizens of Georgia, Alabama, and northern Florida—wonderful people.

Much had changed at Ranger School. Time had marched on, leaving my foggy memories and dreams just that, fantasies. The little cabins at Dahlonega had long since been demolished, though one had been rebuilt to serve as a tiny museum. It was neatly painted, both inside and out; the floor was carpeted; the beds were neatly made with pristine woolen blankets over crisp, white sheets. There were pillows. Good gravy, pillows. I worried for my sanity when I saw it. Could this have truly been the musty, rough-hewn, crumbling structure in which the Rangers had taken refuge from the downpour on the first day in Dahlonega? It seemed impossible. How had my memory failed so completely? The Pentagon had been correct in their assessment that I was a very confused, old man. Perhaps it was time to retire from medicine.

A sergeant major devoted his morning to showing us around the camp, explaining that Ranger cadets were now housed in huge, cinder block barracks, with privacy partitions between the double-decker beds. He brought us inside to show us the thick mattresses, though allowed they had to be replaced every six months, for, "Yeah, times have changed, but Rangers still don't smell too good."

The students were also far better fed than in bygone days. Vanished were the C-rations that had fed millions and millions of GIs over so many decades, those quasi-meals replaced by MREs, "Meals Ready to Eat." There are now dozens of selections, from Vegetarian Lasagna to Chicken Fajitas to Pot Roast. Included in each pouch is a heat pack that soldiers activate with a few drops of water and place next to the foil-wrapped meal. After a few minutes, steaming, tasty meals appear anywhere, anytime—even on the moon. No longer do troops pilfer C-4 plastique explosive from Claymore mines to turn cans of Franks and Beans from congealed cold goop into a mostly-congealed tepid goop.

More shocking, the sergeant major told us of the safety considerations that have been enacted. Dozens of Ranger students have died over the years, often of hypothermia, even at the jungle Training Facility in Florida. Now, behind the scenes, and probably not evident to the Ranger students of today, medical evacuation staff with ambulances, and even helicopters, are in position around the clock to rescue the injured. It was not what I remembered as a priority.

At first, to be honest, the wholesale transformation irritated me. After all, I had held my entire adult life that Ranger training was stark, painful, and smacking with genuine deprivation. It was the only way, I believed, to teach men to drive on in combat when there was no alternative but mission failure and death. These were values that had been set deeply in my consciousness for nearly forty years, and they were the pillars to which I had clung for dear life through my tour in Viet Nam, the numbing years of graduate and medical school, the despondency of residency, a six-month tour with Doctors Without Borders in Darfur Province, Sudan, and three returns to Viet Nam as a volunteer doctor.

The sergeant major, however, sensing my perplexity, set me straight, explaining that the young men who go through the course these days are sent off to Iraq or Afghanistan within thirty days of graduation. He asked, rhetorically, what sense it made to debilitate them immediately before hurling them into the most demanding and dangerous period of their lives.

I also learned that medical studies, done decades before, demonstrated that Rangers suffered significant degradations in health, signaled by changes in blood chemistry at the end of the course, abnormalities which took years to drift back to normal. Coupled with the deprivations of combat, into which many Rangers were sent just weeks out of the course, the results were serious, long-lasting negative health issues for the cream of our military, and, I would argue, the cream of our youth.

Complicated blood irregularities aside, those more recently responsible for designing the course apparently felt that sending young men into combat after dropping forty pounds over nine weeks was so obviously counterproductive, it was tantamount to purposely weakening our troops and threatening success in battle. The seriousness with which these changes were explained, and the solemn commitment of the cadres with whom I had the privilege to speak, convinced me that my initial scorn was misplaced.

The instructors were quick to point out, however, that though the course is now less likely to harm Rangers, it is still designed to push men to discover just what they are made of, and many cadets still do not graduate to bear the Ranger Tab. As in decades past, approximately fifty percent of those who start the course do not make it to the end.

Leaving Dahlonega, we marched for twelve days, often ducking off highways twenty or thirty yards as the sun faded to pass the night in the woods. I covered myself with a poncho and a thin, padded, nylon blanket, a poncho liner, a brilliant invention that ties into the corners of the poncho. It was created forty years too late.

In preparation for the trek, I had done my due diligence as a good Ranger, consulting weather charts for the route, noting the average nightly temperature lows for October and November over the past fifty years. I felt armed with sufficient data to make the trip without dying of exposure and eschewed the thought of toting a heavy sleeping bag. But with Murphy's Law an inescapable ingredient of all projects, particularly those of a fanatical cast, it should not have surprised us that we were confronted with record lows over several nights, awaking encrusted with ice, temperatures on the other side of the meager poncho dipping to 27 degrees Fahrenheit. I remember grumbling to myself at one in the morning, "Hey, knucklehead, you wanted to taste suffering again? Great. Open wide. You're gonna get a full meal of it for the next six hours."

⚔⚔

On day twelve of our odyssey, despite mammoth blisters in places we didn't know we had places, we charged into Camp Darby, the old Harmony Church section of Fort Benning. The ruins of the church are long gone, and the cadres, though vaguely aware of the name, did not know the site of the old building. Most of the top leadership admitted they had not even been born in 1968.

At Camp Darby, named after Brigadier General William O. Darby, who created the Darby's Rangers of World War II fame, we were given a tour of the new facilities. These were sturdy, rugged buildings that will last for many, many years. As in Dahlonega, the students were treated with far greater respect than in the not-so-good-old-days, the emphasis now on imbuing these young men

341

with the hard-boiled skills they will need to carry out their mission, and stay alive halfway around the world, in just a few weeks.

The old obstacle course is long gone, replaced with The Darby Queen, a far more functional and safer set of hurdles. The Ranger course itself is so bursting with military knowledge to master these days, the students delight in traversing the Queen only one time.

The "gulch" over which we had to monkey climb in 1968, and where I lost my first Ranger Buddy, has long since been abandoned and overgrown with trees and brush. The friend with whom I did this march, and who had suffered through my Ranger tales for decades, laughed derisively at the pit, having believed all these years it was truthfully as deep as I had sworn. He did not accept my explanation that it had become filled with nearly half-a-century of silt and flora. The bleachers had also vanished. The stanchions, which I uncovered by digging through the brambles, were just rotted old 4x4s. They were not nearly as far apart as I had remembered the perimeter of the bleachers, and again, my friend was quick to smirk.

⚊⧗⧗⚊

We set off from Fort Benning to march the back roads of southern Alabama and, eventually, those of northern Florida, where the weather turned hot and buggy. We crossed paths with a herpetarium's worth of snakes, most road kill by the time we happened upon them. The nearly three-foot copperhead along a desolate stretch of Poverty Creek Road was the most alarming, even though the creature had long since gone to its final reward. There were countless dog incursions, though just waving our pepper spray canisters at them was usually sufficient to send them off at a scared trot. There were also innumerable deceased armadillos, creatures evidently endowed with a gene that drives them to pitch themselves under the tires of passing cars. All that was left of most of them were sections of their black and white checkered armor plate, scattered like miniature chessboards along the high-way shoulders.

And there were passing cars, probably a quarter of a million of them, most whooshing just two or three feet from us. Why older drivers refuse to pull to the left of an otherwise totally empty road, to grant us more than six inches' clearance from their side view mirrors, is beyond me. With all the time I have on my hands now, without having to walk for nine hours a day, I might apply for a grant from Triple A or Congress to study this phenomenon.

On 11 November 2006, Veterans' Day, after twenty-four days on the road, with nearly 500 miles and close to one million footsteps beaten into our feet, we marched, heads up, shoulders back, into Field Six at Eglin Air Force Base, arriving five minutes early. Here, too, we were greeted with the utmost respect by the Ranger cadres. This facility is now named Camp Rudder, after Major General James E. Rudder, who commanded Rangers in World War Two. The saga of the Rangers scaling the cliffs of Pointe du Hoc, Normandy, France, on D-Day in 1944 is extraordinary. It was the tale related by the old sergeant major in the dunes of Eglin. It might have been better had the old man spoken of that nightmare during the first week of the course. Perhaps we would have more easily understood that while those men were anomalies, the cadres were trying their best to train us to become just as staunch.

The barracks area has been moved to Field Six, a short distance from the site at which we survived three weeks of swamp immersion in 1968, the old Field Seven. The new billets and mess hall are quite impressive in size and sturdiness. The alligator pen boasts many relatively small occupants, one with a gnawed off hind leg. And, on the subject of anatomically incomplete reptiles, there, on the stage of the university-quality lecture hall, sits the stuffed carcass of Big John, her missing left eye and growling countenance somehow preserved. Her innards, however, have been marked and interred in a corner of the field.

The commanding officer was at the gate to greet us. He presented me with a raft paddle inscribed with the Ranger Creed, the date, and the words, "Reliving the Dream." My beautiful wife and precious daughter were there to greet us. I made my daughter, now a United States Marine major, promise that after my wife and I are gone, she would insure that the paddle was passed down to her kids, and then to theirs, and that it would never be abandoned.

# OTHER TITLES BY
# WILLIAM S. GOULD, MD

# CAPTAIN IRON MUSTACHE

*Captain Iron Mustache* takes place in 1968 and 1969. United States Army lieutenant, J.W. Weathersby, just out of Ranger School, volunteers for duty in Viet Nam. It is not long before he changes from naïve youngster to hardened soldier. Something about rural Viet Nam, though, captivates him, and he convinces his commanding officer to allow him to live as the sole American in a remote rice-farming hamlet. His mission is to win the hearts and the minds of the peasants. J.W. forms a deep friendship with the village chief, and falls in love with the schoolteacher, Miss Lin. During a mid-night battle, Miss Lin is arrested and tortured as a communist agent. At the same time, the chief is critically wounded, and disappears after being flown out of the village by an American medevac helicopter.

J.W. and the chief's wife spend the last month of his tour driving the deadly roads of Viet Nam searching hospital after hospital for the man. Nearly half a century later, J.W. and his wife return to Viet Nam in a surreal effort to find the chief. He also wants to see Miss Lin, but the Vietnamese government is suspicious of his motives, and the days of his sojourn are fraught with struggle and frustration until a simple act of kindness changes his life.

# IN BLACK GRANITE

*In Black Granite* is set in the decade after J.W. Weathersby returns from the war in Viet Nam. He eventually accepts the assessment of family, friends, and medical school deans that he will never become a doctor. He drifts without focus until the miracle of his first child's birth rekindles the craving to study medicine. This is a narrative of his dogged struggle to beat the overwhelming odds against a man in his mid-thirties gaining admission to an American college of medicine. *In Black Granite* scrutinizes the ruthless battle for places in medical school, and how the psyches of the chosen are sieved as they are herded through the decade as students and residents. The strain of endless days and nights away from family, of sleepless months, and of pervasive arrogance, distort the souls of even the strongest. Some find the path more treacherous than surviving a war.

# C.O.L.A.

T he day his father died, Dr. Solomon Forte promised his mother he would honor the man's memory by dedicating his years as a doctor to the treatment of injured workers. It seemed so clear a decision—his patients would be like his dad, stoic, honest, working class stiffs who sought nothing more from a doctor than an arm around the shoulder, a word of reassurance, and an ally in dealing with the state industrial insurance system.

His life at the Whitaker Hospital and Medical Center is, though, the antithesis of his dream. He can't tell which of the roadblocks is most daunting: that posed by his medical colleagues, the threats of S.M.A.C., the State Medical Abuse Commission, the bureaucracy at C.O.L.A., the state's Commission on Labor Affairs, or the duplicitous patients, some of whom spend every waking moment trying to dupe him out of drugs and government benefits.

Occasionally, a case is obvious—the worker really was devastated by an industrial accident. It seems to Sol, though, that those are the very patients C.O.L.A. torments. On the other hand, claimants skilled at ripping off the Commission run free for decades. *C.O.L.A.* also examines the specter of serious medical errors, and how they are so much easier to make on patients whose care is mired in the aggravation of government-sponsored insurance plans. Questions are also raised about the state-appointed morality commissions that determine which doctors relinquish their licenses for treating pain. Finally, it is a disturbing look behind the scenes of a modern, multi-specialty medical clinic.

# A HEART WIND FROM
# THE DESERT

D r. Solomon Forte has lost everything. There is little left but to offer himself to the wretched in war-torn Sudan. Arriving in the desert, heart brimming with hope, it does not take long to recognize that the social and political beliefs that have spawned the war and famine are the very forces that prevent him from carrying out his dream of caring for the dispossessed. At first, despite the warnings of the tiny European medical team left at the refugee camp in Darfur Province, he fights back with typical, strident, American resolve to save the entire population of refugees. The obstacles of central African life, however, soon draw the spirit from him, and he turns his efforts to preserving the lives of his Western companions. He falls deeply for a gorgeous, but outwardly hardened, British nurse. When she disappears from camp, he spends what strength is left searching for her. *A Heart Wind from the Desert* examines the need in all of us to accomplish something meaningful in the tiny fragment of time we are allotted, and the impossible hurdles faced when trying to change the way people have thought and behaved for the millennia. It is a tale of beautiful, warm children, but also of the stark life in the sub-Saharan Sahel.

# RAPHAEL'S BLANKET

Raphael Blumenkopf is born clandestinely at the Bergen Belsen Nazi death camp on the 14th of April, 1945. His birth is an unprecedented miracle, as is the liberation of the camp by British forces that very afternoon. He has only his mother and a few surviving villagers from their home in Checzonovska, Poland.

While the majority of the refugees leave Central Europe for Israel and the West, his band travels across Russia to China. A relative has promised jobs in Shang Hai's old Jewish settlement. The journey is fraught with threats from starving Russians, barbaric border guards, and destitute Chinese peasants.

Just as the lives of the immigrants begin to normalize in China, the victory of Mao Zedong's communist army forces them to flee, this time to Hanoi. Five years later, the communist movement in North Viet Nam topples the French government, and the Jews run again. They settle in Saigon until the unrest there compels them to emigrate to America. Raphael's years in the U.S. are colored indelibly by the poison that follows him from the Holocaust, and he formulates a plan to extract revenge from a federal judge with ties to the Nazis. Who could have envisaged the price he'd pay?

# LINCOLN FRIDAY

Lincoln Friday is born into nothing, an obscure, dirt farmer's son, destined to live dominated by the jagged edges of two wars. His early years are an endless series of losses, yet he struggles back after each blow, and slowly, a strongbox of dreams emerges from the fog of his hopelessness.

The harshest test of Lincoln's life, though, comes when the effects of his exposure to Agent Orange devastate both his and his daughter's lives. While the Fridays fight back passionately, the courts, Congress, and the VA turn their backs on them.

In the end, his deeds were neither profound nor dazzling, but he left his mark on disparate people in disparate lands. The world he touched chafed less for his quiet dignity.